GW01393183

WHOSE HEA

So is there a future for the National Health Service? It is now apparent that had we been setting it up today then, irrespective of political ideology, it would have been done differently. As it is, it is with us warts and all, firmly based on the ideas of forty years ago, although it is under increased threat.

While it is also obvious that the demands made upon it are infinite, its funding will always be limited to what it is felt the country can afford – although how high the NHS figures on the list of a government's priorities is very much a matter of political ideology.

What certainly seems to have happened is that although there have been successive changes in the bureaucracy, these have not helped those working at the sharp end. There is no more fat left to cut. As the new managers are now discovering, the ability to run a supermarket chain or command a battleship successfully does not necessarily equip you to run part of what should be a caring organisation where the needs of people should come first.

Also, as we have seen, loud proclamations about the amount of money spent on health, which bear little or no resemblance to what is being experienced in real life, cut little ice. Possibly if those who had so much to say about how splendid everything is actually had to use the service themselves, things might be very different.

Praise for Judith Cook

WHOSE HEALTH IS IT ANYWAY?

'A disturbing piece of investigative journalism
that does little to reassure us that the Health
Service *is* in safe hands'

The Universe

'examines just what has happened to the NHS
and asks questions as to what its future role
should be'

Barnsley Chronicle

'concerned not just with defending the NHS but
also with encouraging changes which would
make it more defensible'

The Health Service Journal

RED ALERT

'has blood-chilling implications'
Books & Bookmen

'a complete history of accidents, incompetence
and deception in the nuclear power industry'
New Statesman

'this chilling, meticulously researched, extremely
important book . . . essential reading!'
She

'For far too long we have been given
deceptions, half-truths, three-quarter-truths . . .
This book looks into some waters that remain
too murky for the public good'
Paddy Ashdown MP

**Also by the same author,
and available from NEL:**

WHO KILLED HILDA MURRELL?
RED ALERT

About the Author

Judith Cook has been a journalist for over 20
years, starting on the *Guardian*. She specialises
in investigative journalism, particularly in social
and environmental issues, and has also spent
some time as a political journalist at
Westminster. In 1982 she was runner-up in the
Campaigning Journalist Award given by the
Periodical Publishers Association and in 1981
she won the Society of Authors Margaret
Rhondda Award. She is the author of THE
PRICE OF FREEDOM, WHO KILLED
HILDA MURRELL? and RED ALERT. Her
journalistic interests also include theatre, about
which she has written a number of books.

Whose Health is it Anyway?

• • •

_____

Judith Cook

NEW ENGLISH LIBRARY
Hodder and Stoughton

Copyright © 1987 by Judith Cook

First published in Great Britain in 1987 by New English Library

First New English Library Paperback edition 1988

This book is sold subject to the condition that it shall not, by way of trade or otherwise, be lent, re-sold, hired out or otherwise circulated without the publisher's prior consent in any form of binding or cover other than that in which it is published and without a similar condition including this condition being imposed on the subsequent purchaser.

No part of this publication may be reproduced or transmitted in any form or by any means, electronically or mechanically, including photocopying, recording or any information storage or retrieval system, without either the prior permission in writing from the publisher or a licence, permitting restricted copying. In the United Kingdom such licences are issued by the Copyright Licensing Agency, 33–34 Alfred Place, London WC1E 7DP.

Printed and bound in Great Britain for Hodder and Stoughton Paperbacks, a division of Hodder and Stoughton Limited, Mill Road, Dunton Green, Sevenoaks, Kent TN13 2YA (Editorial Office: 47 Bedford Square, London WC1B 3DP) by Cox and Wyman Limited, Reading, Berks. Photoset by Rowland Phototypesetting Limited, Bury St Edmunds, Suffolk.

British Library C.I.P.
Cook, Judith, *1933–*
    Whose health is it anyway?—New ed.
    1. Great Britain. National health services
    I. Title
    362.1'0941

    ISBN 0-450-43109-6

Contents

1	Breaking Point	1
2	The Jewel in the Crown	8
3	Safe in Their Hands?	23
4	Lies, Damned Lies and Statistics . . .	39
5	Be Poor and Die Young	58
6	'I Wouldn't Start from Here . . .'	81
7	The Sharp End	94
8	Positive Health	113
9	The Peckham Experiment	122
10	Second Opinions	129
11	AIDS – The Joker in the Pack	151
12	Summing It Up	168
13	Author's Note	173

Notes	179
Select Bibliography	183
Appendix A: The Patients' Charter	187
Appendix B: Inequalities in Health	189
Index	203

Acknowledgments

A lot of people had input into this book. First I must thank Ken Howes, Information Officer of the Association of Community Health Councils for England and Wales for providing most of Chapter 2. Also, Carol Kenway, Information Officer for the group of Manchester Community Health Councils for finding me so many documents on AIDS. I am grateful to Jean Robinson, Tom Richardson and Drs Pietroni and Perrett for their contributions. I should like to fully acknowledge the assistance of the now defunct Health Education Council and thank them for their permission to quote from various leaflets and pamphlets and from their last report, 'The Health Divide', from which Appendix B is taken. Finally I must thank the Association of Community Health Councils for the use of their information and their help and support in writing this book.

1

Breaking Point

The District General Hospital stands just outside the West Country town up a steep hill. It is partly housed in an old workhouse building dating back from the time of Dickens. Outside there is a clutter of prefabricated buildings housing departments like that of pathology – they were put up as 'temporary' accommodation twenty-five years ago.

Inside, the outpatients' waiting room is crowded and stuffy. The resigned queue of patients waits and waits . . . As the morning wears on and appointment times fall more and more behind new patients find there is no longer enough room for them on the hard seats in the waiting room, so they spill out onto the corridors. There is no snack bar and the slot machine for dispensing hot drinks stopped working half way through the morning.

There are a number of mothers with young children and as there is no play area and no toys, there is a continual background wail as they get bored with having nothing to do. This particular outpatients' department has to deal with five times as many patients in a year as it was originally designed for. The paint is peeling, the clock stopped a month ago but, as someone says, at least you don't know how much time you are spending waiting.

As time passes some people actually start to grumble. There are those who have to go back to work, those who have had to rely on uncertain public transport from outlying villages. Deregulation of the buses has meant a much restricted service and they are beginning to worry whether they might have to choose between seeing a consultant or getting home the same day. My time, says a middle-aged man loudly, is as valuable as that of a consultant. People look embarrassed.

The hard-pressed nurses become more and more depressed and helpless. This shows itself in two ways. Either they keep right away out of sight or they walk through the crowd, eyes staring straight ahead, refusing to catch anyone's eye. Doors open and shut, people come and go, still the people outside wait.

The nurses know that there are just too many people to be seen. They also know that some consultants are routinely late, always disorganised and book their clinics badly, with several people given the same appointment time. This is in case one of them does not turn up and wastes the consultant's valuable time. There is nothing the staff can do about it. Nobody can go out and face telling some of the people who have already waited an hour and a half past their appointment time that there might well be further delays.

By lunchtime everybody is thoroughly fed up – patients, nurses, ancillary staff, the consultants themselves. The atmosphere does not make for relaxed consultations or sympathetic treatment. 'Patient needs', that is the need actually to see a consultant, has to take precedence over patient comfort, says the spokesman for the District Health Authority when somebody complains. There are just no funds available to improve the hospital which, as one Health Authority member admits privately, is 'just a rat hole'.

The scene moves to a small manufacturing town in the Midlands where the hospital is equally old but has a new outpatients' department with a far more efficient appointment system. Waiting in the gynaecological clinic is Mrs B. She is forty-five and is having severe menopausal problems.

Her doctor thinks she needs a hysterectomy and she has had to wait five months to see a consultant. He is sympathetic, examines her, and confirms that she does need the operation, but it is not strictly an 'urgent' case. He understands she is very debilitated, that the erratic heavy bleeding is both an inconvenience and an embarrassment, but there it is. She will have to wait for a bed for, oh, say, twelve to eighteen months . . .

Mrs B. thinks enviously of Mrs C. who lives in the same village outside the town. Mrs C. had similar problems. Mrs B.'s husband, however, is unemployed while Mrs C. is the professional breadwinner for her family. She too was told she would have to wait five months to see a gynaecologist, but if she was prepared to pay for a first – and private – consultation then it could be arranged for the following week. Mrs C. wrestled with her conscience and her belief in the NHS and set them against the need to be well enough to continue working and then gave in. She saw the same gynaecologist as Mrs B. He told her, too, that she needed a hysterectomy and he would see what he could do. Nine weeks later, after she had had to be admitted twice for emergency treatment, she was told there was an NHS bed for her. Her initial payment had enabled her to jump the gynaecological queue. She is back at work quite fit while Mrs B. drags herself around trying to cope as best she can. As she leaves the clinic, she passes a newly decorated and renovated gynaecological ward which was opened, with much publicity, by a junior Health Minister some twelve months earlier. Unhappily, it has never been used as the health authority has insufficient staff to man it.

In another part of the outpatients' department Fred is reading the *Sun* and fuming because he has been told not to smoke there. Fred is attending the clinic to see a chest specialist because of his bad cough. He smokes sixty cigarettes a day and, as he frequently tells people, he does not believe in all that rubbish about smoking being bad for you – after all, his grandad smoked sixty Woodbines a day all his life and died in his bed at the age of eighty. So there!

Old Jemima lives in a terraced house in a deprived inner London borough. Her doctor is pretty old too, as are many inner city GPs, and, like them, he works alone, although a workload heavily weighted with elderly and/or deprived people is really too much for him. However, for a variety of reasons – the way the system is organised, the lack of enthusiasm among young doctors for working in such an area, his own stubbornness – he has no partner and uses a deputising service at night and at weekends, which brings its own difficulties. Jemima has a variety of health problems and has an appointment to see a consultant at one of the London teaching hospitals about her arthritis. Because of it she can't use public transport so she has been told to be ready by 8.30 a.m. sharp to be collected by the ambulance. But there have been severe cutbacks to the service and somehow 'her ambulance' fails to arrive at all, so that she waits in all day and finally misses her appointment.

A few streets away old Tom – whom Jemima sees when she manages to struggle to the day centre – is also waiting for an ambulance to take him to the same hospital. He was told to be ready by 8 a.m. although his appointment is at 11.30 a.m. This is because, in order for it to be economic, the ambulance has to collect as many people as possible at one time so, although the journey takes only twenty minutes by car, it takes an hour and a half by the time the driver has roamed the streets looking for patients and has helped the elderly and infirm into his one-man ambulance.

Tom is seen at mid-day, not bad, only half an hour late. However this is only the beginning. All the other people on the ambulance had later times than he did, so he will have to wait until four o'clock before he can be taken home. Finally, the driver rounds up his passengers. It will be two hours before Tom gets home as he is the last to be dropped off and by this time they have hit the rush hour. From leaving home to arriving back has taken ten hours. His consultation took fifteen minutes.

In a near-by block of council flats a young couple who are awaiting the birth of their first child discover that something has gone wrong. Although she is only thirty-two weeks' pregnant, the mother goes into labour and, as planned, is taken to a near-by maternity hospital where she is delivered of a frail, two-pound baby. But there are no facilities for such babies there. It needs a special intensive care unit. Wearily, the doctor who has supervised the delivery starts ringing around to see if she can find an intensive care cot, but is unsuccessful. There are just not enough. Increasingly desperate, she tries hospitals further and further away while the young couple are assured that all is well. The mother worries, as she has not seen her baby since it was taken away after the birth. Finally, one of the hospitals the doctor contacted rings back and says that, after all, they can squeeze the baby in. Mother and baby are rushed off in an ambulance and the baby's life is saved. This particular baby, that is. Another time the doctor might not be so lucky.

In a remote country village Mrs R. is in tears. The family has long since come to terms with the fact that their son, now sixteen, is severely mentally handicapped. When he became too much for his mother to manage he was admitted to the local mental hospital which, although it was old and physically not very suitable, looked after him rather well until he was moved to a newer, custom-built unit housing ten patients. The staff were devoted and kind and Mr and Mrs R. visited their son regularly. Now she has been told that the pleasant unit is to be closed down and her son returned home under the new system of 'care in the community'.

As far as she was concerned he was already being cared for in the community. The Home was in the middle of the town, local youngsters went in regularly to help and make friends with the patients, those able to benefit were taken out on special trips. The staff were kind and sympathetic. Mrs R. has tried to find out what help she can expect for him. A physiotherapist 'should' be able to call once a week, although there can be no promises as the health authority is so overstretched in that department. Has she asked her local council what help it can offer? The social services department can promise nothing. Yes, says a harassed official, they know all about 'care in the community' but there are just no extra resources. Has she tried the DHSS?

Mrs R.'s son is sent home. He is confused and unhappy, having become used to his own surroundings. He can't understand why he can no longer go to a swimming pool or have his hydrotherapy. He misses his friends from the Home. He can't speak too well, so he cannot communicate with the neighbours. He becomes more and more unhappy and disturbed and reverts to incontinence. The social services department says it will do what it can.

It is evening and in countless thousands of homes and in the dayrooms of hospitals, television sets flicker. It is time for the six-o'clock news. A Government spokesman is being interviewed about cuts in the NHS. Cuts? There are no cuts. Never has so much been spent on the health services, the record speaks for itself. New hospitals are springing up all over the place, waiting lists rapidly falling, more patients are being treated at a faster rate than ever before.

The tired people who have waited all morning in the outpatients' department are not impressed. If it is all so wonderful, how come they had to wait so long in such an awful place? Mrs B. isn't impressed either. She remembers the unopened new ward in the hospital. Her neighbour tells her to cheer up, it could be worse, after all Mrs So-and-so was actually admitted for her operation only to find it cancelled for lack of operating theatre time.

Fred puts his fag down in the ashtray as he tucks into an enormous plate of sausages and chips. Later he'll go down to the pub for a few pints. He wonders vaguely what the X-ray will show. I mean, it couldn't be serious – could it?

Old Jemima is very confused and cannot understand why the ambulance never turned up. Tom is so worn out he just makes himself a cold snack and then goes to bed. He supposes he must just put up with things as they are. After all 'they' keep telling him 'they' know best.

The young couple gaze fondly at the tiny scrap of humanity in the intensive baby care unit, blissfully unaware of the fact that it was only a matter of luck that a cot was found in time.

Mrs R. considers walking out on the whole situation. She just cannot cope.

All these stories are based on real-life examples. It looks as if the National Health Service as we know it is at breaking point. What has gone wrong?

2

The Jewel in the Crown

Before we look in detail at the state of the National Health Service today and what direction it might take in the future, it is necessary to look briefly at past history.

Doorstep canvassers during the 1983 General Election campaign speak of their amazement when on mentioning the NHS, some people said brightly, 'Well, there's always been a health service, hasn't there?' and even, in answer to a question as to when it had been set up, 'Under Queen Victoria, wasn't it –' Needless to say these were younger voters who could not remember what it was like before 1948.

Before July 5th, 1948, the day on which the NHS officially came into being, provision had been very sketchy for those who were not in a position to pay the full price for what they needed. Legislation brought in by Lloyd George in 1911 had provided insurance cover limited to manual workers. This entitled them to see a doctor if they were ill themselves, but did not extend to their families.

Those families who could afford to do so paid a small sum weekly to their local doctor, thus going on to what was described as his 'panel', and becoming panel patients. Those who could not afford even this got by as best they could trying to find the money somehow in the event of an emergency which required either a visit to the doctor or – more rarely – calling a doctor out.

If hospital treatment was needed, then there were either

the municipal hospitals or those run by voluntary bodies. Treatment was means-tested and to get a free or very-reduced-payment bed it was necessary to obtain what was described either as 'lines' or 'a recommend'. This was a kind of certificate of need from someone in authority such as a JP, a clergyman or a local councillor. The supply to each was limited.

In 1986 the East Midlands Region of the TUC published a booklet researched by its Health Services Committee into conditions before the setting up of the NHS.[1] It was based on hundreds of responses from those who had experience of what it was like before, and it is worth picking out a few of them just to give the flavour of the times:

> If you needed hospital treatment such as an out-patient and the like you had to have a 'recommend' from your husband or father's place of work. They only had so many and if they had none left, you had to forage around for one if someone had one left.

> My youngest brother was always poorly and needed constant attention. He was in and out of hospital regularly. The problem was getting him there as he had to have a 'recommend' which we had to get from the Chairman of the Council who himself got only about six a year. I dreaded going as he would often say, 'hope you don't need any more this year, I have only one left. They cost 12s.6d., you know.' I later learned that the 12s.6d. was the amount of the voluntary contributions needed to qualify for one 'recommend'.

> Whenever any of us were ill we had to grin and bear it. We four children all had measles, chicken pox and whooping cough. At these times we were all bundled into one room to 'work it off'. A bottle of Ferrin's Fever Cure cured everything.

> The old age pension was ten shillings a week, fifteen shillings for a man and wife, so 2s.6d. was a lot of money

to pay (for a home visit from a doctor). There were a couple of schemes where you paid 3d. a week and the doctor's bills were paid if you were ill. The thing was that 3d. bought a loaf of bread or a few bacon bones to make a broth, and when you were well it was easy to forget that you might need a doctor, so very few workers paid into such a scheme and so the old couldn't afford to do so. That meant people most likely to need care were unable to get it without going into debt.

In 1939 we were living on the 'means test', my husband having been laid off because of an explosion at the Pit. We were receiving the princely sum of 27s. per week. The rent was 8s.2d. Our eldest son was then about 18 months old – he developed measles very badly and the doctor had to come several times a week. It took me months to pay off at a shilling per week which we could ill afford as we had no other source of income.

On reaching the doctor's and after repeated ringing of the bell the doctor finally emerged, took one look at my arm and said 'Yes, it's broken', then he told me to lay my arm as flat as I could on the surgery desk and then he hit it a great wallop with a heavy book and told me to call up the hospital the next morning. This was my complete treatment and now when a doctor sees my wrist and asks how it came to be so misshapen and I explain, they all seem to register disbelief. I have had a deformed right wrist now for fifty years, thanks to a practitioner who didn't consider he was being paid sufficiently to have his Sunday evening disturbed.

And finally:

My husband lost his left leg in November 1935 as a result of a motorcycle accident. While he was still in hospital he was sent a bill for £1 for the ambulance being called. After that he had to pay for his own artificial legs at £25

a time, all repairs, laces, straps, etc. It also cost 4s 6d.
each for pure wool stump socks. Not much now, but as
my husband was an agricultural labourer his wages before his accident were only £1.10s.6d. per week. To
make matters worse, when he was able to get a small
job and we asked the Income Tax Collector if we could
claim the expenses off our tax, a very nice letter came
back in these very words: 'We are very sorry you cannot
claim expenses for your artificial limbs as artificial limbs
are classed as luxuries.' So you can see how grateful we
were when the NHS came in in 1948 and they have been
marvellous to us ever since.

When the Labour Government was elected in 1945 one
of the main planks on which it had fought the General
Election was the provision of a National Health Service.
It was to be the jewel in the crown of the Welfare State,
the best medical treatment for all, free at the point of
need. As well as the best free medical care there would also
be free dentistry, spectacles, artificial limbs and appliances
and prescriptions. It was a brave concept. It was also to
prove very difficult to achieve.

It is not possible in a book of this nature to go into all
the organisations and reorganisations of the NHS carried
out by successive Governments. For most people, at least
until recently, there seemed little connection between
the rearrangement of bureaucracy, and management of
the administrative tiers, and what actually happened at the
local hospital outpatients' clinic or in the doctor's surgery.

Setting up the service in the first place proved far more
intractable than had been anticipated. For a start it meant
sorting out a nationwide system of hospital care from a
mixture of different types of hospitals, all with their own
ownership and managements, the majority of which were
housed in unsuitable old buildings such as ancient work-
houses. Some still are, even in 1987, and there is one
dating back to the reign of George III![2] There were wide
geographical differences in hospital care and widely

differing standards. The 'better' hospitals, such as the big London teaching hospitals, attracted the most highly qualified doctors and specialists.

Although the achievements of Aneurin Bevan as Minister of Health, in piloting the Bill through Parliament and setting up the NHS, cannot be overestimated, some of the problems from which we now suffer date from that time. Not least is that of the position of the doctors. Fearful in the end that he would get no real agreement from the medical profession, he allowed the consultants to dictate their own terms as to how and when they would work within the NHS and how much time they would give to private practice. He left an extraordinary amount of power in their hands. The Act also allowed for GPs to become private contractors to the NHS, not employees of it, setting up the Family Practitioner Committees which administered GPs, dentists, opticians and pharmacists.

Curiously enough for a Government claiming to want to introduce more democracy into the social system by allowing lay representation on various boards and committees, such representation was by appointment, not election, and those sitting on the various health authority boards were not accountable to anybody, a position which still obtains. Because of the amount of time involved this has always meant a preponderance of white, middle-class, middle-aged or elderly people on such committees – the lay membership on all the various health bodies from the Regional and District Health Authorities to the Community Health Councils cannot possibly be described as representing a cross-section of the community.

Right from the beginning the NHS proved to be far more costly than had ever been anticipated, as the Government grappled with the business of updating hospital buildings and commissioning new ones. In the first two years of its existence, Parliament twice had to approve large supplementary estimates to fund the NHS.

But when we look back to those days it is easy to see why those involved could not possibly have foreseen the problems of the next forty years. So-called 'miracle drugs', such as antibiotics, were dealing with diseases which had once seemed intractable. No one then could have imagined that forty years later their continued use would make bacteria resistant to them. Nor was any account taken of the fact that within thirty years of its beginnings there would be a top-heavy ageing population and that the proportion of the elderly would continue to grow.

As to operations like heart and liver transplants, even kidney transplants, these were barely out of the realm of science fiction, while the last killer disease which had terrified entire populations was bubonic plague, the Black Death. It was thought to be the last. No one prophesied AIDS – the Acquired Immune Deficiency Syndrome.

However, with all its problems, for a very long time Britain's National Health Service was the envy of the rest of the industrialised world. Not only was it envied, it was also copied.

So from its beginnings successive Governments have had to grapple with an enormously unwieldy organisation; but, at least until 1979, within a broad consensus that the NHS was A Good Thing. Over the years people's expectations rose, although successive Governments have been fortunate in what seems to be one of the main failings of the British – that many of them seem grateful for anything whether it is good or not and they hate to complain. Because of past history, even today old people can feel that somehow a 'free' health service is charity and cannot be persuaded that, in fact, we have all paid for it.

Escalating costs have always been a nightmare; the first comprehensive review of the NHS came under the Conservative Government of 1951. During the next eighteen years, while Governments came and went, documents and Green and White Papers on the NHS piled up, some to gather dust, others to be partially implemented. When the Conservatives again took office in 1970 the new Health Minister, Sir Keith Joseph, announced that there would be a massive reorganisation of the management structure of the service. His scheme ran into criticism for being too centralised and too management- and cost-efficiency-based at the expense of the needs of the patients, but in 1973 an Act was passed by Parliament to implement it.

The work of reorganisation did not come to an end with the implementation of the 1973 Act. The structure had been changed, but the need remained to manage that structure effectively.

One of the key tasks of management is planning. For the National Health Service, this means ensuring the delivery of appropriate amounts of the right kinds of health care. The unregenerate NHS had tended to develop local services in far too haphazard and uncoordinated a fashion. Planning for change, the capacity to respond to new circumstances and new patterns of demand for services, had not been one of its strong points. From 1974 to 1977, Government policy for the NHS concentrated on rectifying this weakness. A comprehensive planning system was the result.

Area Health Authorities were the institutional foundation of this system. They were conceived primarily as planning bodies, and new mechanisms for planning at all levels – local, regional, and national – followed swiftly in the wake of their creation. The AHAs would develop long-term 'strategic plans', providing a co-ordinated overview of the way money was to be spent across different services and functions. Broad goals would be set for all parts of the local service administered by the District Management Teams. Annual 'operational' plans would show exactly what steps would be taken each year towards the achievement of these goals.

Although the details of these plans would be worked out locally, the overall balance of services was a matter for central government. From now on, decisions about the allocation of resources between different services, such as maternity care or the care of the elderly, would be made in an explicit and systematic way at a national level. Without firm guidance from the centre, it would be all too easy for health authorities to spend money in the same way it had been spent before, without proper consideration of whether this was the 'best' way to spend it. Guidance from the centre was also necessary in order to ensure uniformity of provision across the country. The DHSS would send their circulars down to the health authorities, who would in turn send their plans up to the DHSS.

In 1976, the Government undertook two policy initiatives intended to resolve this question of how to allocate money to different services. They issued two major consultation documents in order to open up a debate and develop a consensus. The more important of these, 'Priorities for Health and Personal Social Services in England',[3] described itself as 'the first attempt that has been made to establish rational and systematic priorities throughout the health and present social services, and . . . the first step in what will be a continuing process'. It proposed a shift in expenditure from the acute hospital sector to primary care and the so-called 'priority groups': the elderly, the mentally ill and the mentally handicapped. The second document, 'Prevention and Health: Everybody's Business',[4] proposed a parallel shift in favour of preventive medicine and health promotion. Both sets of proposals were widely accepted, since it was agreed that for too long the hospital-based acute services had taken a disproportionately large share of NHS expenditure.

As the goal of all this activity was a rational, effective and equitable planning system, it is not surprising that at the same time consideration was given to the distribution of resources between different parts of the country. Should all the Regional Health Authorities receive the same amount of money for each member of their catchment population? Or should they receive differing amounts reflecting their differing needs? The level of provision of services varied from Region to Region. London, in particular, was much better off than the rest of the country; it

had more hospitals, more consultants, and more specialist
services per head of population. The Resource Allocation
Working Party (RAWP) was set up in 1975 to devise a
practicable solution to this problem. It reported in 1976
and came up with a formula for assigning budgets to the
Regions. The effect was to siphon money off from the
'richer' to the 'poorer' parts of the country. Uniformity of
provision was the ultimate goal, though the rate at which
resources would be transferred was left to Ministerial dis-
cretion.

The final piece in this complex planning edifice was put
into place in 1977, when Joint Consultative Committees
(JCCs) were established. Area Health Authorities and
local government authorities shared the same boundaries,
and the development of what is now well known as 'care
in the community' clearly demanded that they co-operate
in planning services for certain client groups, in particular
the elderly, the mentally ill and the mentally handicapped.

These organisational developments took place against a
background of constraints on government expenditure and
widespread industrial unrest in the NHS. The former of
these now seems almost too obvious to mention, so familiar
a part of the political scenery has it become. Had large
amounts of extra money been available at the time, the
NHS need not have concentrated its attention so firmly on
the problem of reallocating scarce resources. This need to
address the problem of scarcity in a systematic way can be
seen as the real heart of the NHS planning system. Services
for neglected parts of the country and neglected groups of
patients could otherwise have been improved by a simple
process of levelling up.

These redistributive policies have not been an unquali-
fied success, even in their own terms. The transfer of
resources from one sector of the NHS to another has
proceeded more slowly than was originally foreseen. The
lack of progress in the care of the community policy in the
1970s is a case in point. It is very difficult to release capital
which is tied up in hospital stock.

The RAWP[5] policy has also attracted a great deal of criticism lately. No one seriously doubts that the goal is a worthy one. On the other hand, many people do doubt whether it is possible to achieve it at a time of low overall growth in NHS expenditure. In this case, the criticism is not that resources are being transferred too slowly, but that they are being transferred too quickly. Patient services are suffering as a consequence. At the beginning of 1987, the Government went some small way to meet these criticisms, but the problem still remains. How much of a burden can be carried by those areas whose provision is better than the average?

The industrial unrest of the 1970s left behind a legacy of a different kind. It was a key factor in starting off a process that was to lead to a revision of the 1974 reorganisation. Between 1972 and 1976 there were three major national disputes in the NHS. In 1972-3, ancillary workers took selective strike action, imposed an overtime ban, and refused to co-operate with management. In 1974, there was a dispute over nurses' pay, and, one year later, the doctors started to work to rule. This quickly became known as the 'pay-bed' dispute. The consultants wanted to practise privately using NHS beds; the Government didn't want them to. Both services and morale suffered, and at the end of the day nobody won.

By 1976 industrial relations in particular and the management of the NHS in general were perceived to be in a very poor state. The problems were compounded by the unexpectedly severe disruption that had followed the reorganisation, and the volume and seriousness of the criticisms directed at the service had grown steadily since 1974. As a result of this, a Royal Commission under the chairmanship of Sir Alec Merrison was appointed in 1976 to 'consider in the interests both of the patients and those who work in the National Health Service the best use and management of the financial and manpower resources of the National Health Service'.

The Commission took evidence and deliberated for

three years. Its final report was placed before Parliament
in July 1979,[6] just a few weeks after a new Conservative
Government was formed under the leadership of Margaret
Thatcher. The evidence the Commission had received
overwhelmingly supported the case for change, and change
was what it recommended. In all, 117 recommendations
were made, amounting to yet another major reorganisation
of the NHS.

There were four central proposals which together would
radically change the shape of the service. First, Area
Health Authorities should be abolished and the smaller
health districts upgraded to the status of full health authorities with an appointed membership. Despite their piv-
otal role in the 1974 reorganisation, AHAs had come
increasingly to be seen as a superfluous and occasionally
obstructive layer of bureaucracy in between the Regional
Health Authorities and the District Management Teams.

Second, Family Practitioner Committees were also to
be abolished, and their responsibilities for the primary care
services transferred to the health authorities. A single and
unified management would administer all local health
care.

'Consensus management' was the name given to the
division of executive responsibility between the members
of a management team; no one person was 'in charge'.
After 1974, the DHSS had issued detailed guidance on how
managers and medical professionals should co-operate in
arriving at decisions. The Commission thought that the
system had failed. Labyrinths of committees and sub-
committees appeared only to delay decisions and absorb
inordinate amounts of time. The system should go – the
third proposal.

Fourth, overall responsibility for the NHS should be
transferred from the Secretary of State to the RHAs. It
was argued that central government had so little effective
control over the service that the Secretary of State's accountability to Parliament made no real sense.

Five months after the publication of the Commission's

report, the Government issued its own proposals in 'Patients First'.[7] It clearly baulked at the extent of the changes proposed. Too large an upheaval would be disruptive. And so, only some of the Commission's recommendations were accepted, though the Government did throw in a few ideas of its own. Consensus management, the responsibility of the Secretary of State, and the separate Family Practitioner Committees were to stay. Area Health Authorities were to go. Comments on 'Patients First' forced the Government to back down on one of its own proposals. They had suggested that the new District Health Authorities would do away with the need for Community Health Councils. With the notable exception of that from the British Medical Association, most of the submissions the Government received strongly favoured retaining the CHCs as 'watchdogs' over the consumers' interests.

In 1981 the required legislative changes passed through Parliament. One year later the second major NHS reorganisation took place and the new District Health Authorities came into being. The powers and responsibilities of the old AHAs were shared out between these newly created bodies and the old RHAs. Districts would now do their own planning within a framework laid down by the RHAs, who would in turn be responsible for monitoring the District's services. Since most Districts didn't share boundaries with local government authorities, however sensible this might seem, the AHA's responsibility for joint planning with local government would be parcelled out to more than one District. Most local authorities found themselves planning with more than one District, and most Districts found themselves planning with more than one local authority. This particular move introduced an extra dimension of complexity into what was already a complex and difficult process, and over the last two or three years the effectiveness of joint planning has been seriously called into question.

A year before this, in 1981, the Government had acted

on a recommendation of the House of Commons Public
Accounts Committee which went some way to meet the
Royal Commission's position on the responsibilities of the
Secretary of State. It took a very different line, however,
from that advocated by the Commission back in 1979.
Instead of formally acknowledging the Secretary of State's
lack of effective management control by transferring
overall responsibility for the NHS to the Regions, it
chose to increase the level of central government super-
vision by setting up 'annual performance reviews'. Each
year the heads of the Regions were to be summoned
to the DHSS to demonstrate their success in meeting
planning targets and implementing central government
policy.

Two years after this, and only one year after the
reorganisation, the Secretary of State appointed Sir Roy
Griffiths, Chairman of Sainsbury's, to lead an enquiry into
the 'effective use of manpower and related resources in
the National Health Service'. It sounded ominously like
the terms of reference given to the Royal Commission.
In many ways the outcome of this enquiry harks back
to the criticisms made by the Commission. The Government was picking up the threads of some unfinished
business.

The Griffiths report made two main recommendations,
both of which were accepted. Consensus management
should be abolished and central government should tighten
up its overall management control by setting up a management board responsible for the implementation of government policy. The process of change is now complete.
'General managers' are in posts throughout the NHS. They
are on short-term contracts and are personally accountable
for the management of units, Districts and Regions. Appointments were often made from industry and the armed
forces, as well as from inside the NHS. Their reappoint-
ment will depend on the quality of their work, as will the
actual level of their pay.

Griffiths' main criticism of NHS management[8] was that

it failed to ask itself, in any systematic way, how well this or that part of the service was doing. There was a need to develop and apply consistent criteria for evaluating the performance of the health services. In essence, this is the task that the new management is being set.

The process of introducing general management has not been painless. The nursing profession in particular has vigorously opposed it, though the original fears of the doctors seem to have been laid to rest. In both cases, there was concern that the new managers would pose a threat to their clinical freedom. Clinical decisions about patient care have budgetary consequences. Similarly, budgetary decisions are bound to affect the clinician's choice between a range of possible courses of treatment. It is easy enough to imagine management pressure to choose the cheapest, rather than the most effective, treatment.

The old-style District Management Teams had represented an institutionalised balance of power between management and the professions. They had formalised the practical requirement of agreement through consultation. Griffiths had argued that this balance of power too often resulted in a stalemate. The channels of consultation had become clogged and sclerotic. What was required was a means of streamlining this process, without disturbing the real balance of power.

It is probably too early to say what the effects of the introduction of general management have been. It was assumed by some that a new broom of efficiency would sweep through the NHS as a result. Others feared an obsession with cost-containment at the expense of patient services. The arguments have by no means ended. To many it seems that power has increasingly slipped away not so much from the clinicians, as from the Health Authority membership. As a result, local accountability is being gradually eroded and replaced by accountability to central government. It should not be surprising if this turned out to be true. Much of the rhetoric of general management has been about the devolution of decision-making. However the underlying tendency of the NHS since the 1974 reorganisation has been towards more centralisation.

3

Safe in Their Hands?

'The National Health Service', trumpeted Margaret Thatcher at the 1982 Conservative Party Conference, 'is safe in our hands.' She was to repeat that constantly during the following eight months' run-up to the next General Election, after which she was returned with a massive Parliamentary majority but only about a third of the vote. Yet even then people had their doubts; in fact it was just about the only issue on which the Labour Party led the Conservatives in the opinion polls. Because somehow it just did not seem safe on the ground, at the sharp end.

By 1987 the disparity between official statistics and the real world seems even more marked. When the row blew up between Norman Tebbit, in his role as chairman of the Conservative Party, and the BBC over its supposed left-wing bias a 'spokesman' for the Party was quoted in the *Guardian* of November 3rd, 1986 as saying:

If you listen to *Casualty* [a BBC soap opera] it is like a Labour Party meeting. The general patois used throughout is so-called health service cuts. We contest this allegation about health service cuts. There are no health service cuts – there is more money and more people [sic]. This allegation of health service cuts is a party political argument but nevertheless in *Casualty* it is used as part of everyday speech.

The Prime Minister has intoned the litany over and over again – NHS spending had increased between 1979 and 1986/7 from £6 billion to £14.8 billion (and in the autumn of 1986 with a General Election in the offing further funds were released). So, after allowing for inflation, Government spending had increased overall by 24 per cent. Between 1980 and 1984 there were 11,000 new hospital beds, 58,000 extra nurses – so what's the problem?

The new breed of managers brought in in the light of the Griffiths Report – some ex-naval or airforce officers as well as some from industry and commerce – were making the NHS ever more efficient although some had, it is true, found the task of running the service as a business too much and had given up. Still, more people than ever were being put through the system, privatisation of services such as cleaning and catering were releasing funds for use elsewhere and more and more people, particularly the elderly, disabled and mentally ill, were being returned to 'care in the community'.

Announcing the publication of the NHS annual report for 1985/6, Norman Fowler, Secretary of State for Social Services, summed up how the Government felt:

> This is the third annual report on the health service in England. Like its predecessors, it tells of record progress made in a service which is of real importance to everyone in the country. In 1985 there were almost 37.5 million hospital out-patient attendances, over 6.75 million in-patient cases and almost 1 million day patient cases. These are all new records and a tribute to the hard work and dedication of health service staff.

There were one or two problems, admitted the Minister. Waiting lists needed to be reduced further, the move towards community care carefully co-ordinated and, of course, there was now AIDS. But 'the health service is prepared to meet these challenges. It is now better financed and better managed. Very significant progress has been

made in replacing outdated hospital buildings and building
new hospitals as a result of the substantial capital building
programme.'

We will look in detail in the next chapter at why Government statistics are not what they seem and what has gone
wrong; but before doing so, let us look at what is happening
from the point of view of those involved – the doctors,
nurses, medical staff, ancillary workers, ambulance
drivers, the Community Health Councils, and organisations such as the Patients' Association and the College of
Health – and, of course, ourselves the patients.

Outside the rosy windows of Whitehall the picture is a
very different one. It is one of rising waiting lists and
insufficient beds, of hospital closures, of ward closures
caused by shortage of staff, of insufficient intensive care
cots for babies, of insufficient beds for cancer and kidney
patients, of deteriorating ambulance services, rocketing
prescription and dental charges, of the kind of incident
where a woman who has waited two years for a hip replacement is admitted to hospital, anaesthetised and then woken
up and told the operation has had to be cancelled, of
elderly people discharged too soon after major surgery
and sent home without anyone to look after them, of the
mentally ill discharged into 'care in the community' and
sleeping rough on the streets.

It would be possible to fill a whole, large book with
detailed and provable examples of all this, area by area,
district by district, hospital by hospital, along with a supplement on what is happening in GP and dentists' surgeries, and to the overstrained social services departments
trying to grapple with those turned over to them from the
NHS. But we can look here at a few real examples at least.

Hospital stories are the most immediately dramatic. A
report published early in 1987[1] explained something of the
position then obtaining in the London area. By January
1987 three-quarters of the acute beds due to be closed in
order to release funds for less privileged areas had, in fact,
been closed. Yet the London health authorities had only

achieved a third of their savings' target. While the number of beds was being run down, the number of patients to be treated was actually going up in part because of the shift in populations towards London from the areas of high unemployment. Only £31 million had been saved by the closure exercise.

Writing in the report, published by the independent King's Fund Centre, Dr John Dunwoody said: 'This means it would be an unrealistic policy to go on closing beds in the hope of hitting the savings target.'

London has four massive Regional Health Authorities and a complex of District Health Authorities. In 1986 alone New End Hospital in Hampstead, the Dreadnought Seamen's Hospital in Greenwich, St Mary's Hospital in the Harrow Road and the St John's Skin Hospital all closed, and proposed closures for 1987 included St James in Balham, Queen Mary's Hospital for Children in Carshalton, Wandle Valley Hospital in Merton, New Cross Hospital and – almost certainly – the world-famous Westminster Children's Hospital. This pattern of closures can, of course, be seen in areas other than London.

Let us select a district like Camberwell, which contains King's College Hospital. In June 1986 King's consultant Dr Roger Williams told a press conference that the King's budget had been cut by £3 million a year for three years and by £1.5 million for 1986/7, resulting in the closure of two wards, cuts in the nursing budget, deferred purchase of vital new equipment and a 20 per cent cut in laboratory tests. Bed occupancy there is 110 per cent.

'On many occasions,' he said, 'a patient has stood in a corridor outside the ward waiting for a nurse to change bed linen which is still warm from the last patient, so that he or she can take that bed.' At Dulwich Hospital a £400,000 shortfall raised the question of kidney patients being turned away or sent to overstretched units at hospitals in other districts (an option not even available in many other parts of the country).

The health district of Lewisham/Southwark is known as

one of the disaster areas of London, with a waiting list that
has risen 15 per cent to 8,243 in the first six months of 1986
and has, in fact, been rising every six months since 1983.
One factor, according to London Health Emergency which
has been monitoring the situation,

> has been the effective closure of Lewisham Hospital to
> all but emergency admissions for much of the year, with
> two wards completely closed in April 1986. Lewisham's
> waiting lists more than doubled in ten months. Six wards
> at Lewisham were closed for six weeks in the summer
> of 1986 and 158 beds and eight out of eleven operating
> theatres at Guy's Hospital were closed during the same
> period.

Guy's kidney unit has been limited to only fifty new
patients a year and sixteen beds have been closed in the
children's heart unit.

In September 1986 the District Health Authority proposed £2 million cuts to meet Government cash targets.
This would mean further ward closures, bed closures, a
three-month freeze on recruiting staff at Guy's, a £335,000
cut from Guy's day-to-day services and £400,000 cut from
schemes to develop 'priority' services for the elderly mentally handicapped; 200 acute beds have been closed at
Guy's since the beginning of 1984, 170 in general medical
and surgical wards and 16 in children's wards. 'The position
is absolutely frightful,' Professor Cyril Chantler, General
Manager of the Acute Unit told the London *Evening
Standard*, complaining that he had been 'asked to reduce
the service below what is reasonable'.

In the West Lambeth District Health Authority cuts
drove waiting lists up 17 per cent in six months in the first
part of 1986. The main hospital, St Thomas's, has had cuts
of some £2.5 million, including the reduction of heart
operations from 400 to 300, a cut of 20 per cent in special
diagnostic investigations and in August 8 wards (197 beds)

were closed to all but emergencies and even to some of those.

There is a point worth making over the August 1986 ward closures at St Thomas's. Cancer is an emotive subject and all those who remember the 1978/79 so called 'winter of discontent', which assisted Mrs Thatcher into power, will remember the television film of cancer patients being turned away from a hospital by striking workers. It made all the newspaper headlines, too. It was one of the most powerful anti-union stories of the whole winter, matched only by that of council grave diggers refusing to dig graves. Together they probably did more to lose Labour the next General Election than any other factor in the dispute.

Yet, in August 1986, cancer patients were being turned away again. On 12 August the front page of the London *Evening Standard*'s late afternoon edition carried a banner headline. It said: 'Cancer Patients Turned Away', and underneath, '79 patients refused emergency admission by St Thomas's'. It went on to explain that shortage of beds was such that there was not even room for cancer patients: 'The hospital lost eight wards at the beginning of this month as part of a package of proposals to cut £2.5 million expenditure in the financial year . . . 79 patients were turned away today, one third of them with confirmed or suspected cancer.'

I was just about to pay for a copy of the paper when it was whisked away, along with the whole remaining pile, and replaced with one with a changed front page showing Mrs Thatcher taking her one week's holiday at Padstow in Cornwall. The massive headline shrieked 'Maggy – Queen of the Beach!' I pursued the earlier edition the length of Charing Cross Road, catching up with it near Tottenham Court Road underground station.

The story did not appear on any of the television news bulletins that night as far as I could see. Nor did it appear in the following morning's newspapers. It was all quite unlike the reaction to the winter of discontent. According to the District Health Authority, among those turned away were two women with breast cancer and one man with a bladder tumour who had had his operation cancelled twice.

St Thomas's was, in fact, cut back to only 107 operations in 67 sessions during that month and there was also a cut of 59 patients at a drug addiction clinic there – what was all that we heard about the fight against drug addiction? The specialist St John's Skin Hospital closed permanently, along with four wards at Tooting Bec Hospital. At the time of writing another money-saving plan is to close radiotherapy in-patients' services at Guy's and King's College Hospitals and focus all in-patients on the already overstretched St Thomas's.

In Greater London 7,767 hospital beds have been permanently closed and not replaced since 1980 (this includes 10 per cent of acute beds) and 32 hospitals have been closed, 3 partially closed, 21 approved for closure and 5 approved for partial closure. The closures of acute beds brought the first 'Red Alert' (i.e. emergency admissions only) since the 1973 cold weather 'flu epidemic. Waiting lists in the London health authorities are up 15 per cent overall from the peak of the health workers' pay dispute of 1982 – from 108,000 in 1982 to 118,000 in 1986. Bloomsbury's waiting list alone is up to 15,000 – the population of a small town.

The story of hospital and ward closures in London can be repeated from Coventry to Cornwall and Norwich to Cardiff. More than 36,000 beds, 10 per cent of the total, have been cut since 1979. While new beds have come on stream, proportionately more have been closed and there are instances where whole new wards remain empty and locked up because there is insufficient money to pay nursing staff.

Dr Donald Menzies, consultant at Liverpool's Women's Hospital and chairman of the medical executive committee of the District Health Authority, told the *Observer* on February 1st, 1987 that while the number of consultants in his department had been reduced from 4 to 3 the throughput of patients had increased by 30 per cent, leading to increasing complaints and overstrained staff. 'Patients are staying in hospital for shorter times. For Caesarean sections, they used to stay in for a fortnight and now it is eight days. For hysterectomies it used to be three weeks, now it is also eight days. But you get patients readmitted because of complications.'

But if the big cities can blame their plight on RAWP how about the rural areas which were supposed to benefit from it? Cornwall is a case in point.[2] It is a long, thin peninsula with no large town or city and it has always suffered from its distance from where the decisions are made, along with lack of resources. The South West Regional Health Authority area is sprawling and cumbersome, taking in Gloucestershire through to Cornwall, with funds administered from Bristol. Average incomes are among the lowest in the country and many patients requiring specialist treatment, especially for cancer, have to travel right outside the county to Plymouth in Devon.

It has a relatively new District General Hospital at Treliske in Truro and an old City Hospital housed in a Dickensian workhouse. Many local cottage hospitals and wards in smaller hospitals have been closed and many services centralised in Truro. Patients come in from a catchment area of some fifty miles. Treliske Hospital is now so pressured that for a year it has been on *permanent* Red Alert. As in London, patients have had to be put into beds still warm from their last occupants and patients have been sent home so early that readmissions are, to quote a senior member of staff, 'no longer infrequent'. Emergency admissions can spend up to four hours on a stretcher in the corridor while staff desperately seek for a bed.

The position became so acute that in January 1987 senior consultant physician Dr Michael Winterton turned to the media, pointing out that Treliske was one of the busiest hospitals in the whole country, with the fastest turn around of patients, and that patients were suffering and staff were under an intolerable strain.

He was prepared, he said, to make it an election issue

– almost an unprecedented statement for a senior medical man – because 'I feel this has gone on long enough. We're one of the busiest units in the country yet we are always being left out of the reckoning when we come to funds . . . Cornwall deserves a proper hospital. If any Government cares about Cornwall and the NHS in Cornwall, then it should be prepared to make up forty years' deficiency and give us a proper hospital to do a proper job.' Staff were considering having to send more patients to Plymouth – over seventy miles away.

Every aspect of the health service is suffering, including preventive medicine. Far too many women in Britain still die of cervical cancer. Our record is poor compared to, say, Scandinavia. For many years there was a great drive to get women to go for smear tests so that the condition could be diagnosed at its earliest stage and successfully treated. Women who had had a smear went on to an automatic recall system run nationally. One of the economies made during the first Thatcher administration was to stop the automatic recall system. The onus was put on either local health authorities or doctors to call in patients or on the women to ask for a test. The result was what anyone might have expected. Cases of cervical cancer began to rise.

In 1985 Oxfordshire Community Health Council went to press with stories of the death of one woman, and the serious illness of two others, following failure to recall them after they had had positive smear tests. A survey carried out in the wake of the publicity this aroused showed that only 7 out of the 201 health authorities had a proper recall system and a total of seventy-seven had no scheme at all. Doctors wanted a national – and this time computerised – recall system set up at an estimated cost of £16 million but the then Minister of Health, Kenneth Clarke, rejected this out of hand.

In what has become all too typical a response there was much pious talk about the gravity of the situation, all Regional Health Authorities were asked to set up their

own systems and to improve the effectiveness of the laboratories which processed smears – but there would be *no*
extra funding available for this.

In November 1986 *Pulse* magazine said that Social Services Secretary Norman Fowler had been sitting on an
unpublished report on cervical cancer which had been
delivered to him some five months earlier. It warned
Ministers that immediate intervention was necessary to
save lives. The report, prepared by the DHSS's own statis-
ticians, covered the period 1974/83 and showed a higher
incidence of cervical cancer in the north than in the south
– another north/south divide – with 'unacceptable' levels
in South Tees, North West Durham, and North Tyneside.
In East Cumbria, Northumberland and Sunderland the
incidence was higher than average and rising fast. The
three-month wait now usual for the result of smear tests
was 'unacceptable', said a doctor commenting on the *Pulse*
story. At the time of writing the report remains unpublished.

But even if you actually are told the result of your test
fairly promptly, this is no longer the end of the story. By
December 1986 women in Cornwall were waiting up to
five months after receiving a positive result before they
could even see a gynaecologist unless they were prepared
to pay privately. One woman who had such a wait had
already had a mastectomy for breast cancer – she finally
saw a doctor in North Devon – and another who waited
five months to see a consultant then waited a further two
for an exploratory operation and a further four after she
had been told a hysterectomy was urgently necessary.[3]

Another speciality which can also be preventive is adequate nursing of premature and sick babies. It can not only
save life it can save severe handicap. Yet a *Guardian*
survey published in July 1986[4] showed that this service too
is on the point of breakdown.

It told the story of a premature baby delivered at Epsom
Hospital which began to bleed internally and needed surgery and intensive care. The nearest hospital with the

facilities was at Carshalton (it has since been marked for closure) but it had no spare cot. The consultant in charge spent four hours on the telephone trying to find a cot and finally one was offered at the Great Ormond Street Hospital for Sick Children in central London. The baby died shortly after arrival, following its long wait and long ambulance ride. 'Perhaps the baby would have died anyway,' said the consultant in charge, 'but the circumstances certainly reduced its chances for survival.'

Dr Andrew Whitelaw of Hammersmith Hospital told the *Guardian* that in the preceding twelve months he had been unable to admit three babies; two had died and one was now severely mentally handicapped. A doctor at the West Middlesex Hospital had tried to find a cot for a sick premature baby and it had taken three days, by which time the baby was also severely handicapped. At that point in time twelve of the thirty-four cots at University College Hospital could not be used owing to a shortage of nurses and the department was turning away more babies than it could take.

Nor is the problem confined to London; in fact, outside it is even worse. Dr Malcolm Chiswick of St Mary's Hospital, Manchester, said a survey of his own on the fate of babies he had been unable to admit showed a death rate 30 per cent greater than for other babies. In Bristol, Dr Peter Dunn at Southmead Hospital said his baby unit had had to be closed to outside referrals, leaving many hospitals in the south west stranded. Babies near to death had had to be sent to Cardiff, Birmingham and even Liverpool, from Bristol and also from districts much further west.

Some 3.5 million working days are now lost in hospital outpatients' queues, said a report from the National Consumer Council in August 1986.[5] British people are passive queuers but their self-discipline was being abused, said the chairman of the Council. Clinic sessions were overbooked and too busy. Too few consultants were seeing too many patients.

In 1986 the Association of Community Health Councils

looked at the results of surveys carried out by some fifty-five of its members and this confirmed the findings of the NCC.[6] From Chorley in Lancashire to Cambridge city, from City and Hackney Community Health Council to Cornwall, the situation was bleak. Lengthy waits of one to two hours were commonplace, overcrowded and uncomfortable waiting rooms with no adequate facilities the norm. People did not like to complain. At the City Hospital in Truro 65,000 people a year were put through facilities designed for 15,000, there was often insufficient seating and there was literally nowhere to park in an area where most people had to come by car owing to the lack of public transport.

The surveys found too that, contrary to popular belief, under 10 per cent of patients attending hospital outpatients' departments were taken there by ambulance. The ambulance service too are now suffering nationwide.

London Ambulance Service had a shortfall in funding of £1.8 million in 1986, equivalent to an under-provision of 233 staff. The immediate effect was felt on non-emergency services, where there was a marked reduction. There have been increased cancellations of outpatient journeys, outpatients being brought in and collected up to three hours late, ambulance patients being left behind because they were not immediately ready to be taken home and the discharge and transfer of elderly patients taking place in the evenings.

Once again the Community Health Councils began turning up stories across the country of similar reductions in the ambulance services; indeed, in some rural areas even the emergency services are no longer functioning properly. The national criterion for the arrival of an ambulance to an emergency call-out is that it should arrive within twenty minutes of the first call 95 per cent of the time. By the winter of 1986/7 this was down in some rural areas to percentages in the mid-sixties, and some organisations were looking at the legal position of their health authority to see if cuts in the ambulance service (under whatever

name, i.e. 'reorganisation', 'two-tier') could represent a
failure of duty under the NHS Acts.

Hospital and ward closures, cut-backs in ambulance
services, longer waits for hospital admission are all part of
the cost-effectiveness and savings which are supposed to
have been made. Sometimes the figures stand between us
and the real people they represent – such as the old people
who die of old age before they reach the top of the list for
a hip operation.

Another way savings were to be made was by putting
out various services to private contractors, e.g. cleaning,
catering and laundry. By 1984 COHSE, the Confederation
of Health Service Employees, had discovered that some
thirty-five Conservative MPs had rushed to take up consult-
ancies with, or directorships of, companies set up to do
this work. Nor was it only MPs who were involved – the
Prime Minister's husband, Denis, is also connected with
one of them.

Horror stories of filthy wards and appalling food and
lack of even basic laundry have become commonplace in
local and national newspapers. One excellent example of
what can happen was discovered by Riverside Community
Health Council when two of its members paid a visit to
the maternity unit of the Westminster Hospital in June
1986. It is worth quoting fairly fully from it as it is typical
of what has been found elsewhere.

The Riverside Health Authority in London, a huge
sprawling district recently combined from two previous
health authorities, had put its cleaning work out to a firm
of private contractors in the summer of 1985 – a firm which
had already been criticised by a number of other health
authorities who had been or were using it.

When members of the CHC visited the maternity unit
they found it had only three baths for thirty women, one
of which was unusable as a dirty mop had been left in
it along with cleaning equipment. The one shower,
and its surround, were thick with built-up dirt and the
shower head was virtually blocked. The washbasin was

blocked and a midwife said it had been like that for five weeks.

Sinks and draining boards in the kitchens where food was served were 'in a disgusting state and encrusted with long term dirt and grease'. One sink was full of dirty dishes and the floor in 'a sordid state'. The sink and draining boards in the treatment room were 'filthy' and the floors dirty. One ward, recently closed for cleaning, was thick with dust. In another ward a patient said the floor had been cleaned only once during her stay and in the antenatal clinic pieces of glass lay around on the floor for two days. The floors in the unit were only vacuumed and no longer washed.

The patients' day room was 'disgusting', with dirty cups and saucers lying on tables along with pools of foul liquid. Ashtrays overflowed with cigarette ends, window ledges and tables had hardened stains. In an alcove off the end of the day room a cleaning woman had been found sleeping at 7.30 a.m., a few weeks earlier. She had put several chairs together to lie on and told patients in the day room not to put on the TV as it might disturb her. When she was discovered she retorted: 'What do you expect when I'm only paid £70 per week.' Cleaners earned £2.10p. an hour with no bonuses or overtime. The previous in-house rate was £3.

Midwives told the Community Health Council that they spent a considerable amount of time chasing up cleaning staff or doing it themselves and this imposed a great burden and ate into the time they could give to patients. One ward sister said cleaners were not even given adequate tools and each was only issued with one jug of diluted washing up liquid per day.

In a statement issued later Riverside Community Health Council went so far as to advise mothers not to go into the unit to have their babies if they could make other arrangements. It said the cut-price cleaning policy was a false economy for real costs increased as trained professional staff had to waste their time on it, while patients

were subjected to cross-infection and other health hazards.
As for the cleaning staff: 'They must come from the bottom
of the labour pool. They do not receive any meaningful
training to do their job and they are poorly supplied with
cleaning materials. In short, they are treated like the dirt
they are supposed to clean.'

We will look in the next chapter on 'care in the community', where savings are supposed to be made on the
NHS Bill by efficient care of long-stay patients back in
their own communities. A worthwhile notion with which
few could quarrel – so long as adequate funds are made
available. It requires careful co-operation between the
NHS and local authority services and it is not in reality a
cheap option.

Among those sent back to the community are many of
the long-term mentally ill. While no one could support the
continuation of the vast Victorian lunatic asylums which
still house far too many of our mentally ill, in some areas
cut-backs have caused the closure of even small purpose-
built units right in the communities they are supposed to
serve.

For far too many of the mentally ill, turned out of
hospital, there is nothing at all. Relatives cannot or will
not look after them – and some of them have no relatives.
Owners of inferior boarding houses are prepared to make
a profit taking some of them in, others have no alternative
but to sleep out. For those who live outside London
the sight of muttering, dirty, obviously disturbed people
travelling around on underground trains or camping out
on the roadside at night is shocking. One can only presume,
from the apathy with which they are viewed, that a large
proportion of the public now accepts the situation as normal.

What happens to them in Britain in the twentieth
century, in our society which has never had it better,
with our NHS on which so much is supposed to be being
spent?

On February 2nd, 1987 there was a letter in the *Guardian*

from three doctors, Dr Peter Silverstone, Dr Conor Duggan and Dr Alain Gregoire, all of the Department of Psychiatry of King's College Hospital.

In it they said that it was now recognised that up to 25 per cent of all homeless people in London are mentally ill:

> That these people may suffer unduly from the effects of the cold weather is illustrated by three patients with psychiatric disorders whom we have seen in the past two weeks.
>
> The first of these patients, a thirty-two-year-old man, developed severe frostbite necessitating the amputation of both legs. In the second case, a patient set fire to a mattress to try and keep warm having developed frostbite, thus causing burns to his feet. In the third case, a seventy-year-old lady failed to seek any shelter, developing severe hypothermia and is still seriously ill. Alcohol was not a factor in any of these cases.

For people in this state an extra £5 allowance for heating was obviously irrelevant. 'Surely,' ask the doctors, 'we can do more?'

4

Lies, Damned Lies and Statistics . . .

So why is it that the National Health Service seems to be in the state it is in? And why is there such a discrepancy between what the present Government (at the time of writing) says is happening, and the money that is apparently being spent, and what is obviously happening on the ground?

Well, to begin with, from being the model nation on which so many others based their later versions of a health service we are now at the bottom of the heap. We are the smallest health spender among the Western developed nations and although the amount spent on the NHS in 1984 was, on paper, a record it was in fact the smallest annual rise in growth for ten years. The proportion of gross national product the UK devotes to health has lagged behind the Organisation of Economic Co-operation and Development average by 25 per cent per year. In 1960 the UK was one of the nine largest health spenders in the developed world. Now we are right at the bottom, while the USA, Sweden and West Germany spend 40 per cent more of their gross national product on health than we do.[1]

There are a number of reasons why the amounts spent by Government on the NHS are not as impressive as they sound. First and foremost is the fact that the rate of inflation within the NHS has risen far more rapidly than the general rate of inflation for a whole variety of reasons. New techniques of a costly nature have become more and more commonplace – heart transplants are a good example of this. There is an increasing proportion of elderly people compared to the population as a whole. Unemployment and poverty have increased the strain on the NHS (whatever government spokesmen may say) and we will look at this aspect in the next chapter; and none of this really faces up to what is surely going to be an appalling problem by the end of the decade – that of AIDS.

When the amount actually spent on the NHS is more carefully analysed, then it is possible to see that there are wide variations even within the service. An excellent report published in 1986 by the Institute of Health Service Management, the British Medical Association and the Royal College of Nurses[2] showed that the rise in spending available to the hospital and community health service had been much less than that for the NHS as a whole and during some years actually fell.

The most recent figures showed that there had been four years up to 1985/6 with very little increase in revenue spending or none at all.

In 1985 the three bodies who commissioned the report just mentioned had calculated that 'in the light of the multiple objectives set for the NHS, a 2 per cent per year increase in real spending for the next three years is the minimum required for the General Managers to be able to carry out their tasks'.

Yet what do the most recent figures for the last four years up to mid-1986 show? They show that the growth rate in spending for the NHS as a whole, deflated by pay and price increases, grew by only 0.6 per cent in 1985/6. In the area of hospital and community services, the growth over the four years up to mid-1986 was, respectively, 0.8 per cent, 0.0 per cent (nil growth), -0.1 per cent (actually *minus* growth) and 0.4 per cent – all to be set against the 2 per cent absolute minimum calculated by those actually working in the service.[3]

The response from Government to the need for a minimum 2 per cent increase in real terms was that more account must be taken of efficiency savings in generating resources for new programmes but, as Nicholas Bosanquet of the Centre for Health Economics who worked on the 1986 report, points out, there are now real difficulties in relying so heavily on efficiency savings to provide the necessary development funding for the NHS.

The report identified several areas which were absolutely crucial when looking at future funding and among these was the need to adjust for demographic change. In 1984/5 the number of people over the age of seventy-four rose by 40,000, bringing the total rise over the period of 1979-85 to 300,000 or 13 per cent. As those over seventy-five are obviously heavier users of the health service than, say, people in their mid-twenties, the inference is obvious, but it does not appear to have been properly taken into account. Also in 1982/3 the spending per head on hospital and community health services was nine times greater for a person aged over seventy-five than on somebody between the ages of sixteen and sixty-four, so spending should actually increase to meet these additional needs. Yet between 1982/3 and 1984/5 expenditure on the age-sensitive services actually *declined* by 0.1 per cent, while spending on some other services increased by only 1.8 per cent which is still inadequate.

In fact the overall picture of expenditure on hospital and community health services takes us back again to the King's Fund report mentioned in the previous chapter, with cuts of £109 million required in Inner London health districts between 1983/4 and 1993/4 (12 per cent), equivalent in real terms to the combined annual costs of St Thomas's, St Bartholomew's and the Royal Free Hospitals; a reduction of between 7 and 31 per cent in each district's spending on local acute services and an overall reduction of 15.7 per cent in local acute beds.

This had all been geared to an anticipated decline in the number of admissions to Inner London hospitals of 15 per cent during that ten-year period. But admissions have not declined. On the contrary they have actually increased by

2.5 per cent, leaving a greater number of people chasing a diminishing number of beds.

Another area where the facts don't seem to fit is in the area of nursing. Every time there is a story about a ward closing for lack of nursing staff or the removal of beds or intensive care cots due to staff shortages the reply comes back promptly from the Government that there are more nurses than ever and that the Government has actually paid for the NHS to take on 58,000 new nurses during its term of office.

But this is not what it seems. The figure of 58,000 'extra' nurses is based on the reduction of the working week from which the numbers of 'full time equivalent nurses' is calculated. In 1980, the basic working week of nurses and midwives was reduced from 40 to 37½ hours. This magically 'increased' the whole-time equivalent workforce by 7 per cent without a single extra hour being worked or a single full-time extra nurse employed. The figure of 58,000 also masks the number of part-time nurses employed instead of full-time nurses.[4]

The position is especially serious in specific types of unit. Take the case of intensive baby care units. Highly trained nurses with specialist knowledge in both midwifery and specialist intensive baby care are paid only £6,500 (1986/7) for this demanding job. In many instances, too, the DHSS has decreed that hospital ancillary buildings – including nurses' homes – must be sold off. In the spring of 1986 the DHSS asked all health authorities to draw up plans to sell 'spare' properties, including nurses' homes, flats and houses. This should raise around £750 million nationwide. Some Regional Health Authorities, such as South East Thames, were already doing this in order to try and raise funds and had sold over 250 properties between 1985 and spring 1986 at a cost of £7 million.

This has meant that thousands of nurses, already among the country's lowest-paid workers, are facing having to find accommodation on the overpriced, overstretched housing market, and where prices in London and the south east especially are now grotesquely high. This will obviously affect nurses working in units such as baby care and in the London area most of the intensive baby care units are now understaffed by about 50 per cent.

Nursing babies in such a unit requires minute-by-minute monitoring and the Maternity Alliance (an umbrella organisation made up of members of a wide range of voluntary groups interested in mother/child welfare) estimates that there is a shortfall of 40 per cent in the number of intensive baby care cots needed in Britain. Some would put it even higher than that. Of the cots available, only 9 per cent have a staffing ratio greater than 1 nurse per cot although the recommended ratio is 4:1.

There are countless other examples of shortage of nurses of which I will mention only two. In January 1987 Charing Cross Hospital in London had to close its specialist coronary care unit indefinitely because there was only one trained nurse left there out of a complement of ten. Patients had to be sent out on to general wards where it was more difficult to provide expert care and monitoring.

At that time the hospital had 149 nursing vacancies out of an acute establishment of 671. In the *Guardian* of January 7th, 1987 Miss Aileen Phillips, the hospital's Director of Nursing, was quoted as saying:

> The shortage is getting worse, particularly in specialist areas like coronary care where extra training is needed and nurses become disillusioned if they do not feel that their work is properly appreciated.
>
> We have been relying on agency nurses for over a year to keep the coronary unit open – up to half the nurses have been coming from such agencies – but now we are down to one staff nurse and the situation is just not possible.

In a note to doctors at the hospital Dr Roland Guy explained that other patients might have to be moved to free beds where heart patients could be monitored and that junior doctors would have at the end of each shift to hand on care trolleys to the next team. 'If you feel that a patient has suffered because of lack of adequate facilities, I suggest you record this in the notes and inform your consultant,' he advised, 'and your protection society and the unit manager.'

Problems of staffing are particularly acute in the Cinderella service of them all – mental nursing in psychiatric hospitals. In 1986 a private enquiry was held in the Trent Regional Health Authority into a series of seven suicides of people under the care of one particular hospital in its region, all of which took place in the period between March and May 1986. Among these was a woman who hanged herself on a near-by golf course, a man who hanged himself in one of the wards, a woman who went home and then killed herself and another patient who jumped into the River Trent and drowned.[5]

The unions of the ancillary workers, such as the National Union of Public Employees (NUPE) and the Confederation of Health Service Employees (COHSE), spoke of increased levels of violence at the hospital at night, especially when there were too few trained staff on duty.

Shortage of staff made it difficult for nurses to take sufficient account of the state of each patient. 'You can't actively watch suicidal people if you have other things to do,' said a COHSE spokesman. 'You need someone to talk them out of it. The staff are very worried that it's easier for an enquiry to blame them rather than the wider considerations.'

Probably there is no area of greater contention than that of hospital waiting lists. The Government, while admitting that there is a problem, sees them diminishing and promises a 'blitz' will be mounted to bring them down.

Yet even here sources differ. The Office of Health Economics in its Report for 1986[6] says that waiting lists rose by 20 per cent between 1975 and 1984 and now stand at 817,000. The Government's 'official' waiting list figure, given in a written answer to a Parliamentary question and reported in Hansard (September 30th, 1986), is 681,900.

In October 1986 the *British Medical Journal* summed up how many doctors feel in an article headed 'DHHS Waiting Lists – A Major Deception?' The study took as its starting point the premise that waiting list statistics for England and Wales contain many anomalies and as a result this meant that several categories of patients awaiting treatment were excluded. (This could explain the discrepancy in the figures given above.) So the study concentrated on one district general hospital in Manchester which seemed to be typical of the national picture and analysed the lists of patients waiting for admission.

The DHSS circular SBH 203 instructs staff to exclude various categories of patient when calculating waiting lists. This exclusion covers day cases, those who wish to defer admission for personal reasons, patients who have already failed to accept an offered date for admission and those who cannot be admitted earlier for medical reasons, e.g. patients asked to lose weight before undergoing surgery or those needing special treatment for cancer.

Altogether 2,094 patients under the care of ten consultants working were checked. When the figures had been sifted thoroughly and the patients taken out who had defaulted when called for admission, it seems that at this hospital the true number of patients awaiting admission was a massive 79.9 per cent greater than the official figure and that the magnitude of the difference varied from one surgical speciality to another.

Long waiting lists do, in fact, affect every area of the country. For example, 78,000 people were trapped in waiting lists in the West Midlands by December 1986 and it is beginning to look to some of those involved in their care as if many of them will now never be treated. The area is one of the worst in the country and includes those waiting in what one doctor described as 'the never never queue', awaiting surgery for either varicose veins or cataracts.[7]

The facts emerged during a survey conducted by the

West Midlands Health Management Centre. It discovered that some lists stretched back an appalling seven years and delays of more than two years could mean in some cases that treatment was never carried out. One of the researchers involved in the survey later told the local paper that if you took the twenty worst districts for waiting list problems in the country then the West Midlands had five of them, with Coventry as one of the country's worst black spots. The situation had become so dire that he was suggesting radical solutions were needed – such as contracting out surgery to private hospitals.

One way the Thatcher Government has measured success within the NHS is by 'throughput' of patients; that is, the number of patients going in and out of hospital. More patients are being treated than ever before, they say, by fewer medical and nursing staff; so, runs the argument, this must be more cost-effective. Staff numbers in the NHS overall have, in fact, fallen from 829,000 in 1982 to 816,000 in 1985, while the number of in-patients treated has risen by 8.8 per cent.

But as the Institute of Health Service Management report mentioned earlier points out:

> The usual measure of productivity of throughput rather than output is in terms of the numbers of cases treated, i.e. a measure of activity of throughput rather than output. A rise in the number of patients being treated may not be a sign that the health outcomes have improved. Indeed, it may simply mean that services have to adopt a revolving door policy. Furthermore these measures will be biased as the emphasis shifts further towards primacy care and day treatment. Much more information is required about re-admissions.

This is certainly true. From Cornwall to Merseyside taking in Inner London, there is proof that readmissions are on the increase. But this does not appear in Government statistics for the simple reason that each readmission is counted for DHSS purposes as a *new* admission. So the readmissions actually boost the number of patients going through the system to help make it appear that a greater number of them are being treated than ever before. It is thus almost impossible to calculate how much money is actually wasted by discharging patients too soon and then having to readmit them when they develop complications outside that they might never have developed had they stayed long enough in hospital in the first place.

Much of this organisation is now in the hands of the new breed of managers brought in in the wake of the Griffiths report. The DHSS leaned heavily on the health authorities to appoint from outside even if they preferred someone who had made a career in the health service. Retired military personnel seem to be popular and there are a number of ex-wing commanders and ex-submariners now running sections of the health service. Some have not been able to stand the heat in the kitchen.

In the area which covered the psychiatric hospital mentioned earlier, the Trent Regional Health Authority, the RHS settled for an ex-Signals colonel in preference to the incumbent professional administrator with wide NHS experience. After only twenty months in his job he resigned over the terms of his contract and his departure followed that of no fewer than five second-tier managers and that of the head of the mental illness unit's community care programme. This, too, is a pattern repeated nationwide as the new breed of managers discover that it is just not possible to run sections of the NHS like a private business.[8]

But, of course, the NHS covers many other areas, it covers family doctors, dentists, opticians, chemists, community nurses, midwives, health visitors, occupational and speech therapists and many, many more. Far more people become involved with the NHS at this level than at that of hospital admission.

According to government figures funds to family doctors
have increased by about 31 per cent, although after ac-
counting for pay rises and the average rate of inflation for
drugs (which is much higher than the general average rate
of inflation) this reduces to about 21.2 per cent; but it is
substantial all the same. Family doctors are not Regional
Health Authority employees – they are private contractors
who contract out their services to the NHS.

The Family Practitioner Committees administer the activities of the private contractors – doctors, opticians,
dentists and chemists. Until 1984 the FPCs were part
of the health authorities, but they are now completely
autonomous health authorities in their own right.

But while general practices in some parts of the country
have been burgeoning, there are very real problems in
others, such as those in the deprived inner city areas. These
areas have the highest proportion of single person practices
(i.e. just one doctor), old doctors and poor facilities. They
tend to make more use of the deputising services in part
because of being single-handed and the use of deputising
services nationwide has come in for a good deal of criticism,
although it is obviously a useful service if properly
used.

For those who live in deprived inner city areas just
getting on to a doctor's list at all can be very difficult. In
passing it is interesting to note that, according to the
General Medical Services Committee, although the number of general practitioners has increased in recent years,
average list size – a measure of the ratio of GPs to population – has only recently fallen to the levels prevailing in
the 1950s! Compared to other developed nations, the UK
has a low ratio of doctors to population and expenditure
on all doctors in the UK is a relatively low proportion of
total health service costs (12 per cent) compared with other
industrialised countries (14-19 per cent).

To return to the problems facing patients in inner cities,
the trend has been apparent for some time. In 1982, for
example, Bloomsbury Community Health Council looked

into the difficulties of finding a doctor in its area. Four people phoned every GP practice listed there – a total of sixty-two. They posed as an elderly male pensioner, a young unemployed male, a young working woman and a young mother with two children.[9]

They discovered that there was a hard core of practices in the Westminster part of its area that flatly refused to take on *any* new patients at all for any reason. Of the remainder it appeared that it was slightly easier for an elderly man to be accepted than a young mother of two. Yet in many cases doctors had specifically chosen to work in single-handed practices. That all these people, and especially the elderly, 'should face such difficulties is really inexcusable when one considers that GPs who work in single-handed practices by choice should, in theory, be prepared to take on all groups of patients,' said the report.

Certainly one of the most expensive areas of the NHS is that of drugs. In 1984/5 the cost of drugs to the NHS was £1,562.50 million. Yet when the Government was given an opportunity to cut the drugs bill by means of generic prescribing (as suggested by the Greenfield Report of 1982) it bowed to the pressure from the massive pharmaceutical industry. ('Generic' prescribing meant the prescribing of the basic form of a specific drug rather than one of its more expensive brand-name versions. Generic prescribing would still have left doctors with a choice of over 600 drugs they could prescribe without seeking any special permission.)

Instead, in 1984, the then Health Minister, Kenneth Clarke, announced that there would be a 'restricted list' of basic drugs which could be supplied on the NHS, any alternatives having to be paid for on the basis of their full cost. Amazingly, a list of only thirty drugs was drawn up as being adequate for basic needs for an NHS prescription; for any of the others, the doctors would have to seek special permission.

There was an immediate outcry from doctors and health professionals, who rightly pointed out that the list was inadequate. (Missed off, for example, were the special laxatives for terminal cancer patients, who were expected to use the most old-fashioned version, which was quite inappropriate.) Doctors asked who had decided which drugs should be on the list and what criteria had been used.

But this information was far too sensitive to be revealed. Believe it or not, it apparently came within the bounds of 'national security', as all those involved in deciding what went on the list had had to sign the Official Secrets Act! When the Minister was asked in the House of Commons on February 1st, 1985 why this was necessary, he replied that the request to sign it was 'more or less automatic'. However, later on, after increased pressure, the list was enlarged. But it still falls between two stools, neither cutting costs while still giving doctors a varied choice, as recommended in the Greenfield Report (which would have left about 600 basic drugs prescribable under the NHS), nor giving them the total freedom which had obtained before and which certainly suited the drugs companies. As long as they fill in the necessary form doctors can still prescribe the most expensive version of a drug if they feel they can justify using it.

But it is at the sharp end that patients feel the cost of prescriptions. Again and again, after every rise, the Health Minister of the day assures everyone that only about a fifth of the population actually has to pay as there are so many exemptions. But even if this were strictly the case, look at what has happened.

When Mrs Thatcher came to power in 1979 the cost of a prescription was 20p. an item. She immediately increased this to 45p. – a rise of 125 per cent. In the following years the charges were 70p. (1980), a rise of 56 per cent; £1 (1981), a rise of 43 per cent; £1.30 (1982), a rise of 30 per cent; £1.40 (1983), 8 per cent; £1.60 (1984), 14 per cent; and £2 (1985), 25 per cent. At the time of writing (early 1987) the charge is £2.20 (10 per cent) and this is shortly to increase to £2.40. It is a truly massive increase and there is growing evidence that the cost of prescriptions is most definitely deterring patients from seeking medical help.[10]

Prescriptions are now seven times more expensive in *real* terms than they were in 1979. The Government has defended this on the grounds that 'those who can afford to do so should make a small contribution to the government's increased spending on the NHS', but there seems to be a shortage of relatively affluent patients and the number of prescriptions paid for has fallen rapidly. Between 1979 and 1984 it was down 35 per cent in England, from 107 million to 70 million.

One Welsh GP wrote to the *British Medical Journal* with the results of his own small survey conducted in his rural practice. Half those patients issued with a prescription for which they had had to pay said they had put off going to see the doctor because of the cost. Half also said they would have to go without something in order to afford the drug prescribed – most by spending less on food. This pattern must obtain elsewhere.

The General Medical Services Committee says in its 1986 Report:[11]

> We consider the present arrangements for prescription charges to be inequitable and anomalous. The 1946 NHS Act introduced a free and comprehensive health service. The 1949 Amending Act allowed for the introduction of prescription charges and these were implemented for the first time in 1952 and continued until 1965. Charges were reintroduced in 1986, with exemption for the old, young, chronic sick and those on low incomes. In addition, contraceptive drugs and appliances were exempted from the time of their availability on the NHS in 1975.

The GMSC goes on to say that the present system places a burden on only one in five; those charged are not necessarily those able to pay, and so the charge is now at a level which deters some patients from seeking treatment.

Exemptions for the chronic sick apply to a relatively short list of well defined conditions 'but the list is incomplete and illogical'. Contraceptive drugs and appliances are now excluded and this, too, is illogical. Exemptions on grounds of age (both young and old) remove large numbers from the category of those charged without any consideration of ability to pay. The wholesale exemption means that the level of prescription charges is much higher than it would otherwise have to be to raise the same revenue. In fact the level of charge is such that it can often exceed the cost of the prescription, which means that a patient can actually be prevented from purchasing a medicine privately from a pharmacist or dispensing doctor if its cost is less than the charge that would otherwise be levied.

As with prescriptions, so with dentists, for dental charges too have rocketed up. This concerns many dentists, who see years of educational work persuading patients to come for regular check-ups and to seek preventive dental treatment going down the drain as charges prevent patients from seeking help in time. There is anecdotal evidence now of people asking for teeth to be extracted rather than filled in order to save future cost – something which had almost been eradicated. In January 1979, before the Thatcher Government came in, the cost for a course of treatment for those not entitled to it free was £5. This did not include dentures, crowns, inlays and gold fillings. Charges for these were £10-£15 for one, two or three teeth, depending on the material used, up to £20-£30 for more than one denture. Crowns cost £10 per tooth, with a maximum of £30 for three teeth or more.

By July 1986 the total cost for a course of treatment was £115, although 'a small amount of treatment might only cost you £17 or a little less'. The cost of dentures or bridges for one to three teeth had risen from £10-£15 to £26-£50, the total for a denture or bridge from £20-£30 to £47-£98.[12]

During the second Thatcher term the optical service was privatised and by July 1987 it will be completely so, only sight tests remaining free on the NHS. Everyone in need of spectacles will have to buy them in the private sector and only children and the very poor will get any help with the cost and that will be in the form of vouchers. At the time of writing the Government had not even decided if those with the poorest sight who need complex and expensive lenses can even be included in the voucher system.

Looking at all three fields, when you read government statistics think of these, quoted in *New Society* of March 14th, 1986: 'In the dental service, the most expensive course of treatment cost roughly 20 per cent of the average manual wage in 1976. Today, it is around 70 per cent. In 1979, the then 20p prescription charge accounted for less than a tenth of the average cost of each prescription. Today it is almost half.'

Finally, a topic to which we will be returning. It was during the second Thatcher term that the concept of 'care in the community' really got under way. No one argued that for many people such care would be the best. There have been enough reports of a grim nature on the conditions inside, for example, long-term mental institutions, which contained many people who with help, could live a more fruitful life outside.

Elderly and disabled people were taking up much-needed hospital beds. So they should be returned to their own communities, where they could be cared for in familiar surroundings. But it was stressed by everyone – those working in the NHS, in local authorities whose social services would be involved, by the voluntary organisations on whom a burden was found to fall and, not least, by the families of those involved (the majority of carers being women) – that if such a scheme was to work properly it could not be a cheap option.

There would need to be more money, not less. Finance would have to be sorted out between the NHS and the local authorities; there would need to be extra community nurses and local authority staff, money for adapting public and private premises, and, above all, proper organisation. Little of this has happened and concern is growing that the situation is rapidly getting out of hand.

For some old people the option was a private residential or nursing home and the Government encouraged the setting up of such homes by offering large subsidies for each patient. Standards varied enormously from the appalling to the good. Only a handful of inspectors are employed to police the homes to see what the standards are and there is a very real fear that if the Government decides at some point to cut back on its subsidies and there is an unexpected boom in the tourist trade, residential homes will revert to being guest houses and old people will find themselves with nowhere to go.

Even the Audit Commission in a draft report published in September 1986 speaks of the disarray and confusion in the community care service. 'It is no longer sufficient to muddle through, hoping that a community care service will emerge,' it told the Government. 'The government must rationalise its funding policies and provide short term funds to reduce the transition period, otherwise two inadequate services will be struggling along in parallel indefinitely.' It went on to speak of agencies pulling in opposite directions and, overall, a danger of the most vulnerable people being left without care and without hope. The extra cost of the service is likely to be millions of pounds, bridging finance is desperately needed for the short term, with the short term getting longer and longer. Where bridging money is inadequate, community care is starved of funds and there are just not enough trained staff for it to work. The mechanism of shifting funds is still inadequate, leaving huge sums locked up in hospitals while local authorities cannot expand community services for fear of being rate-capped. Some of those local authorities constantly chastised by the Government as high spenders are the very ones struggling to cope adequately with community care. It is a vicious circle.

The National Audit Commission suggested that there should be a new single body, separate from the NHS and local authorities, which would be responsible for assessing the needs of people in community care and prescribing how they should be met. It would have control of its own budget and would 'buy in' the necessary services.

But at the time of writing this is not being accepted and the situation gets worse by the week. This is particularly so in the field of mental health and a November issue of SHELTER's magazine, *Roof*, quotes Richard Parker, branch secretary of COHSE at St John's Hospital, Lincoln, where, as in Trent there had been suicides: some fifteen current or former psychiatric patients had killed themselves in the eighteen months up to November 1986. He said: 'The dream of community care is becoming a nightmare.' Workers at St John's say that a system which was meant to provide more humane ways of treating the mentally ill has turned into one which merely shunts them, uncared for, into communities that don't want them. A Salvation Army hostel warden described the situation to SHELTER as 'horrendous'.

Former patients without any network of community support were being placed in multi-occupied slums where landlords claimed full board for patients from the DHSS while giving them one meal a day, or where they took the Giros from patients but gave them no money back. Some patients were placed in small, mixed-sex, multi-occupied bedrooms where they were turned out into the streets between 9 a.m. and 5 p.m. each day. One landlord, claimed SHELTER, was making £1 million a year from the DHSS payments from registered residential homes to accommodate the mentally ill.

It would appear that day centres are overwhelmed, and the families to whom some of those returned for care in the community have been sent have cracked up or broken down under the strain. For some people there are no homes at all, no community care. They sleep in parks, on pavements and on the beaches. At the time of writing

there has been no impartial evaluation of 'community
care', how it is working (or not working, as appears to be
the case), what it is costing and how much is needed to
make it work properly. As it is, more often than not it
appears to remove people from NHS statistics but does
not remove the problems.

So it is possible to show that the glowing statistics so
often quoted at us are not always what they seem and that
other statistics tell a different story.

But probably one of the most exasperating decisions
made by the Government, which certainly upset many
people working within the NHS, was to offer financial
incentives to the new managers to make changes which
would cut costs.

The DHSS was pretty coy about the circular which set
out this notion (DHSS circular PM (86)7) and it was hard
to get hold of a copy. But the effort was worth while.
District managers, said the DHSS, might miss out on their
annual pay rise if they failed to achieve individual targets
under the new merit scheme. Those who worked within the
scheme and consistently exceeded short-term objectives,
making 'excellent progress towards long-term goals',
would receive extra percentage awards on top of their
normal pay rise. In effect, they could expect bonuses of
up to £3,960 according to the extent of the changes they
were expected to achieve – and top of the list of 'savings'
came ward closures. Other suggestions included increasing
the items put through the laundry from, say, 20,000 to
25,000 without increasing unit costs, and identifying
areas for improved efficiency and maintaining financial
control.

So there we have it. It will actually pay the new breed
of managers to close down wards and take beds out of
service, irrespective of need. This will do a number of
things. It will, indeed, cut costs – financial costs that is,
at the expense of the human cost. It will improve the
Government's own brand of statistics and enable it to
demonstrate how well the NHS is doing, how much more

efficient it is and how more people are being treated in fewer wards and fewer beds. And the lucky managers can expect about £4,000 extra for providing the Government with such a service.

5

Be Poor and Die Young

What we have been discussing might well be called the National Health Service but what it is, in effect, is a National Sickness Service. It is increasingly geared to deal with what happens when things go wrong. We take ourselves to the doctor when we feel ill, we go into hospital when something goes wrong but in reality defining our health and that of the nation is not as simple as that.

Where and how we live and the way we live all play their part in how healthy we are, and there has been ample evidence for a long time now that as a nation we are in a pretty poor state. The health of the average person living in Britain today is less good than it is in most other developed countries and this appears to be getting worse, not better. Not surprisingly, the poorer you are the worse your health and the less chance you have of seeing old age.

First let us look at 'Charter for Action – Health for All by the Year 2000'. This document was prepared by the Faculty of Community Medicine in this country for the World Health Organisation. As well as outlining the responsibilities of everybody concerned with the pursuit of better health in Britain, to which we shall be returning, it produced some horrifying information, not least that Britain has some of the highest rates of death, disease and handicap in the whole of the developed world.

To quote from the document:

– the number of deaths from heart disease is not falling as it is in other developed countries and we now have the highest death rate in the world from that cause;

– the expectation of further life at age 45 is among the worst in the developed world;

– infant mortality over the last twenty-five years has declined less than in most European countries and continues to show marked differences among regions of the country, social classes and major ethnic groups;

– the number of deaths from cervical cancer has hardly changed in the past fifteen years although numbers have halved in other European countries;

– the number of deaths from lung cancer among women now rivals that from breast cancer, despite knowledge of how this disease could largely be prevented;

– alcohol intoxication is associated with nearly half the road accidents involving pedestrians, of fatal accidents involving drivers and of all severe head injuries.

This is for the whole population.

There are three broad reasons for the generally poor standard of health in Britain today. The first is deprivation, the second a combination of ignorance and lack of effective health education and the third, and possibly the most contentious, is environmental.

There has been no shortage of studies, reports and surveys on the effects of poverty and deprivation on health but probably the most important of these was published in 1980. During its term of office the 1974-9 Labour Government set up a working party under the chairmanship of Sir Douglas Black to study the effects of deprivation on health and to compare differences in standards of health between the different social classes.

It reported on its findings in 1980 but by that time there was a problem – there was a new Government. The

Conservatives had come to power in the June of 1979. The Report was entitled 'Inequalities in Health Care' and you might be forgiven for never having heard of it or for having heard only of something vaguely referred to as 'the Black Report'. Using a ploy which was to become more common as the years went by, the new Government 'published' it (if that is the correct word) on August Bank Holiday Monday. It was never properly printed, some 250 photo-statted typescripts being made available to a select few.

To ensure that no one would be in any doubt as to the Government's stance on the Black Report, the Secretary of State for Social Services wrote a foreword for it in which he stated that he did not propose to respond to any of its highly detailed recommendations. A copy of the original report is a rare, if not priceless, document but later Penguin Books were to publish a version of it in paperback with the poverty expert Professor Peter Townsend as editor.

No Government has wanted to admit the links between poverty and poor health, least of all Mrs Thatcher's. The Black Report found that all the most recent data considered by the Working Party showed marked differences in mortality rates between the occupational classes for both sexes and at all ages.

At birth and in the first month of life twice as many babies born to the families of unskilled workers in Class V died than babies born to Class I professional class parents, and during the next eleven months of life four times as many girls and five times as many boys died in Class V as in Class I. In later childhood the ratio of deaths between the different classes fell to 1.5:2.0 but increased again in early adult life.

While Black noted that a class gradient could be observed for most causes of death, it was particularly steep in the case of respiratory diseases and the available data on chronic sickness paralleled that of disease.

It is possible here only to look very briefly at some of its findings but they are important. Self-reported rates of long-standing illness were twice as high among unskilled manual workers and two and a half times as high among their wives, as among the equivalent professional people. Black pointed out that the extent of the problem is illustrated by the fact that if the mortality rate of Class I had been applied to Classes IV and V during the period of 1970-2, then the lives of 74,000 people under the age of seventy-five would have been saved. This estimate included 10,000 children and 32,000 men of working age.

Inequalities also existed in the use made of the health services and, 'most worryingly' said Black, the preventive services. Those at the bottom end of the scale used them least, partly because insufficient services were provided in deprived areas and partly because of the cost of using them, i.e. transport, taking time off work and other such practical difficulties.

There has been a good deal of evidence since in support of Black. In 1983 Professor Peter Townsend, speaking at a conference in London ('Patients' Needs First') spoke of work being undertaken by his unit at Bristol University which fully supported the findings of the Black Report. A study of two electoral wards in Bristol, one wealthy and one deprived, showed a mortality ratio of 2:1, from stillbirths right through to deaths among the elderly.

A report in the *British Medical Journal* in December 1984[1] showed that the mortality rate in the UK is higher than anywhere else in Europe. The study, undertaken by John Catford and Sherry Ford, showed that the ratio applied to deaths due to circulatory and respiratory diseases, heart disease and cancer. In Scotland twice as many men and almost as many women died of heart disease than in Belgium, Denmark, France, Greece, West Germany, The Netherlands, Norway and Sweden.

When Professor Townsend spoke he pointed to the factors he felt contributed largely to the poor health of those in low incomes. He cited unemployment – then standing at around 2½ million – the numbers living on or below the poverty line, poor housing and poor nutrition. Professor Townsend rightly pointed out that all

Governments have contributed to the present state of affairs but that the Thatcher administration had deliberately moved away from the consensus politics with regard to the welfare state which had obtained since the end of the Second World War.

There is no doubt that, at the time of writing, there are far more people living in deprived circumstances and it is likely that this will be showing up in health statistics for a considerable time. Unemployment has risen from 1,665,000 in 1980 to – using Government figures – just over 3 million in 1987. (Many experts put that figure as being nearer to 4 million. The Thatcher Government has altered the way in which unemployment is calculated nineteen times since 1979!) Hundreds of thousands of other people are involved in a whole variety of government schemes which are not only low paid, but by no means ensure a regular job afterwards.

Spokesmen for the Government from the Prime Minister down deny that unemployment has any real effect on health, but a recent study of the serious effects of unemployment on health by the Nuffield Centre for Health Service Studies, in Leeds, backs up a growing body of evidence that unemployment effects both mental and physical health not only among those who are unemployed themselves, but within their families.[2]

It showed that the unemployed suffered from higher rates of states of anxiety and depression than those in work, with the rate of illness going up in direct proportion to the length of time out of work. In fact an earlier study undertaken in Edinburgh between the years 1968 and 1982 showed that suicide rates in that city had jumped between ten and twenty times compared with those for people in work and in 1982 the unemployed accounted for 58 per cent of the suicides in the city.

A study in 1986 for the Government's Office of Population Censuses and Surveys by Professor John Fox of the City University, London, showed that unemployed men were twice as likely to commit suicide as those in employment, 80 per cent more likely to have a fatal accident and 75 per cent more likely to die of lung cancer. Death rates among unemployed men in general were 21 per cent higher than expected in every age group and death rates among their wives were 20 per cent higher.

But it is not only the unemployed who are at risk, but those caught in the poverty trap for other reasons as well, such as those in low-paid work and single parent families. By 1983 (the last year for which adequate statistics are available) 16.4 million people in Britain were living on, or beneath, the poverty line. Of these, according to the Low Pay Unit, 8.9 million were living on or below supplementary benefit level and nearly 3 million had fallen through what is described as the poverty 'safety net'. This represents an increase of one third since 1979 in the numbers below supplementary benefit and nearly half in the numbers on or below supplementary benefit.[3]

This quite obviously has an effect on health. A study of social security claimants showed that 50 per cent of couples on supplementary benefit with children ran out of money part-way through each week. 56 per cent were in debt, 60 per cent lacked a complete standard set of clothes, 3 million people could not afford to heat their homes properly (and deaths from hypothermia among the elderly were running at nearly 100 a week during the winter of 1985/6), 6 million people lacked an essential item of clothing such as a winter coat and 5.75 million people have to go without food on occasion through lack of money.

At the time of writing a couple on unemployment benefit receives £49.80 a week to cover all their needs – that is, £3.55 a day each. Couples on supplementary benefit receive £48.40 plus housing costs and get £10.10 a week – or £1.44 a day to cover the cost of a child under ten, i.e. for everything – food, clothing, heat, etc.

The poorly paid and unemployed are particularly vulnerable where housing is concerned. Poor housing can bring a whole range of health problems with it, particularly respiratory diseases. The present Government has cut spending on housing some 60 per cent since 1979 and since 1983 the proportion of the national income spent on housing in the UK has been the smallest in Europe. In 1982 Britain spent 2.1 per cent of its gross national product on housing compared with an average of 5 per cent by all the other OECD countries.

Low-price housing is an essential for those on low incomes but only 33,600 council houses were started in 1985 compared to 174,000 in 1975. Councils, forced to sell off their stocks of council housing, have also been forbidden to use most of the funds raised by so doing to provide new council housing. Housing subsidy has been cut to less than a fifth of what it was in 1979 and a Department of the Environment survey in the autumn of 1985 showed that council estates need to have around £20 billion spent on them for essential repairs. The amount needed to put all the housing stock into a state which makes it fit to live in is far, far larger.[4]

Between 1983 and 1986 a wide range of benefit cuts were implemented, among which was a cut of over £250 million from housing benefits, cuts in single payments for furniture and other essential needs and the abolition of child dependency additions paid as well as short-term National Insurance benefits.

The new Social Security Act of 1986 will mean still further cuts in benefits, with some 3.8 million people losing from the reforms made under the Act, including 2.2 million pensioners. Under the Act additional weekly payments for special needs such as heating or particular diets will be abolished and the new Income Support Scheme will not make up fully for their loss. The value of the State Earnings Related Pension will be cut by a half and the very poorest people, on supplementary benefit, will now have to pay a proportion of their rates. Many rights to appeal against social security decisions will be weakened or abolished.

To add to the army of those likely to produce future health problems are the 100,000 homeless families – and that figure is a conservative estimate (in both senses of the word . . .). There are 1 million people on council house waiting lists. Those who are now homeless include many young people who have been forced to keep moving on because of Government restrictions on board and lodging payments and who often end up sleeping rough, families in 'temporary' bed and breakfast accommodation and those unfortunates who were persuaded into buying their own council houses and then found they could not manage the repayments when, say, the breadwinner became unemployed. If these people lost their homes through defaulting on their mortgages, even if this was unavoidable, then they are classed as 'intentionally homeless' and cannot be rehoused by the council.

Even school meals, virtually a lifesaver for poorer families for many years, can no longer be relied upon to give school-age children at least one nourishing meal a day. By 1980 the Government had raised the cost of school meals from 10p. to 35p. Later in that year they brought in the Education Act which allowed local authorities to charge what they liked and even scrap the school meals service altogether. Children entitled to free school meals often find now that only sandwiches are provided for them and even where school meals are cooked the nutritional value is often very low. Many school canteens have gone over to a cafeteria system offering chips with everything.[5]

That the Government is becoming sensitive to the country's poor health statistics was shown in August 1986 by an event which had echoes of the publication of the Black Report on August Bank Holiday 1980. Every ten years the Registrar General produces a supplement to the national census in which deaths are analysed by occupation, cause of death, sex and age. The last one was slipped out in August 1986.

The *British Medical Journal*[6] asked why the document had had such a low profile and also why reference to the social class differences in mortality had almost been removed. The previous supplement, said the *BMJ* under the headline 'Lies, Damned Lies and Suppressed

Statistics', had been twice as long as the new one and a
sixth of the price. It had also had sixty pages devoted to
discussing social class differences in mortality.

The new one had 128 pages of commentary, a set of 87
microfiches and only 5 pages were devoted to the subject
of death and social class. The way the supplement was
presented – although there was a wealth of statistics on
the microfiche – made it very difficult to extract useful
information, but it did show that the health gap between
rich and poor had sharpened dramatically in the four years
between 1979 and 1982. Death rates among young and
semi-skilled and unskilled workers aged between 25 and
44 were twice as high as those for professionals and managers of the same age. Women in social classes IV and V
had suffered too and were a staggering 70 per cent more
likely to die than wives of men in Classes I and II.

One of the things Professor Peter Townsend had to say
at that 1983 conference is just as apposite now as it was
then, or more so. He said that Government health plans
had not seriously addressed trends in health and the social
distribution of poor health and disability in our society.
No Government, he said, should be allowed to get away
with such irrational conduct.

One of the recommendations of the Black Report had
been the setting up of a Health Development Council
through which attention could be called to the problems
of poverty and health 'which the present government does
not appear to want to ventilate. There is inventiveness and
expertise lying wasted in the country which should be
used to provide a running, but dignified, commentary
on the inadequacy of current government reasoning on
health.'

Concern for public health should

cut right across party lines, it is something shared by the
honourable tradition of the Tory party, upheld by
'the late' Lord Stockton, by the SDP, the Liberals and
the Labour party. It is concern about the polarisation of

a nation not reflected in most other capitalist societies. Indeed, in most other countries contrary measures have been adopted. I think the situation is so serious that we have to talk about it in very drastic terms because the inevitable deterioration of the health of our nation is the direct consequence of the enlargement of poverty in our midst.

On March 24th, 1987 there was a virtual re-run of what happened when the Black Report was published. The Health Education Council had commissioned an update on Black from an established health researcher, Margaret Whitehead, to look into inequalities in health care in the 1980s.

In the last week of its existence, the HEC made arrangements to launch its findings with a press conference. A couple of hours before the launch the chairman of the HEC, Sir Brian Bailey, who will also be chairing the Thatcher Government's new non-independent body which replaced the HEC, tried to impose a ban. The HEC and Sir Douglas Black sidestepped the ban and the document was launched at a hastily convened press conference in a room behind a guitar shop in Shaftesbury Avenue! The HEC printed 2,500 copies and by the lunchtime of March 25th they had all gone . . .

The new Report was even more damning than Black. It showed that although things were pretty bad when Black reported, they had got steadily worse since 1979. The Report notes that the then Health Minister, Patrick Jenkin, when he received the Black Report had said that the cost of meeting its recommendations was 'quite unrealistic in present or any foreseeable economic circumstances, quite apart from any judgement that may be formed of the effectiveness of such expenditure in dealing with the problems identified'.

The findings of the new Report show that whether social position is measured by occupational class, or by assets such as house and car ownership, or by employment status, a similar picture emerges. 'Those at the bottom of the social scale have much higher death rates than those at the top. This applies at every stage of life, through to adulthood and well into old age.' (See the graphs in Appendix B.)

All the major killer diseases now affect the poorest occupational classes and the unemployed more than the rich. Working class people are shorter, more liable to obesity, more prone to disease, with higher blood pressure, lower survival rates for cancer and coronary heart disease and have worse teeth. The diseases of 'affluence' have vanished.

The unemployed and their families have considerably worse physical and mental health than those in work. 'Until recently,' says the Report, 'direct evidence that unemployment caused this poorer health was not available. Now there is substantial evidence of unemployment causing a deterioration in mental health which improves on re-employment.'

Women still have lower death rates than men but experience higher sickness levels. However, 'recent studies have shown that the situation is far more complex. Women's health varies with social class, employment and marital status in ways which are only just beginning to be understood. The health of working class women is particularly poor.'

When considering access to health services by different groups, there is evidence of poorer provision in the more deprived areas and poorer quality 'although no overall picture of the extent of the inequality around the country is available'.

The Report paints a depressing and very worrying picture. When it turns to what action has been taken to implement the recommendations made in the earlier Black Report, the answer appears to be – none. 'The response to the recommendations has been characterised by a lack of action particularly at national and central level. No unifying lead has been taken to guide or motivate a response which would have translated recommendations into action.'

In fact some of the services that Black wanted to see revitalised have actually deteriorated. For example, provision of day care for under fives, and the school health services, which were ways of giving children a better start in life, are all now giving cause for concern. Similarly, with services for the disabled the shift in resources towards support for people in their own homes has been minimal. Prevention and health education are still not being given the priority that was recommended.

Black had seen the abolition of child poverty as the number one priority for the 1980s. Yet more and more children are falling into the poverty trap. In 1971 22 per cent of the poorest fifth of the population consisted of families with children, but by 1982 the figure had reached 30 per cent:

If poverty is defined as income marginally above the level of supplementary benefit, then in 1981 over 3.5 million children in the UK were living in poverty. The situation continues to deteriorate. By 1983 the Child Poverty Action Group estimated that one third of all children living in Britain were living in families in or on the margins of poverty. In 1983 400,000 children were living in extreme poverty, below the supplementary benefit line. Coupled with those on the supplementary benefit level, this represented a doubling of numbers since 1973.

In 1979 a total of 11.5 million people were experiencing poverty. In 1981 the figure had increased to over 14 million (27.5 per cent of the population). By 1983 it had increased to 16.3 million (30.5 per cent of the population). During that period other indicators of deprivation showed similar increases. The number of homeless people increased from 40,000 in 1979 to 140,000 in 1984. All these figures have continued to grow, relentlessly.

Conversely, the Report shows that as more and more people have fallen into poverty in the 1980s, the rich have got richer, after decades of a gradual closing of the wealth gap. According to the Inland Revenue, the share of marketable wealth owned by the richest 1 per cent of the population had been dropping steadily since the war until 1980. In that year the richest 1 per cent owned 20 per cent of the total wealth. By 1981 it had risen to 21 per cent, where it stayed until 1984, when it began to rise again. Trends show it still rising and this is attributed to, among other things, a boom in the stock market, the easing of certain capital taxes and cuts in income tax, linked with substantially bigger pay increases for those already on high incomes. 'All these factors,' concludes the Report, 'have benefited those who were already wealthy, while doing little to help those who were poor.'

Finally, to pick out just one more example, the situation regarding nutrition has deteriorated steeply.

The school meal service, for instance, envisaged by Black as a source of free nutritionally sound meals for all children, is in jeopardy in many local authorities. Fees have had to be increased, the nutritional standards were abolished altogether in 1980. Many authorities have introduced self-service fast-food arrangements, and financial constraints mean that in some areas the service may disappear altogether.

As the *Observer* said, when commenting on the Report: 'The connection between the Government's economic policies of the past eight years and the nation's health could hardly be more evident.'

Unless you are totally locked into what are popularly called Victorian values you are unlikely to believe that all poverty and deprivation is self-inflicted and that people become unemployed, live in appalling housing conditions and fail to keep themselves warm and well fed by choice.

But, equally, it is apparent that some of the nation's ill health is self-inflicted. We all know old So-and-so who smoked sixty cigarettes a day, drank eight pints and several Scotches every evening, lived off chips and died at the age of ninety . . . or if we don't then there is always someone who does. But for most people cigarettes, alcohol and a poor diet are not a good thing. They do lead to a definite effect on health and are a burden not only to those who suffer, but also on the National Health Service which has to cope with the results.

The Royal College of Physicians estimates that at least 100,000 deaths in Britain each year are caused through smoking.[7] It is an interesting figure to consider as we begin to ponder the problems we are about to be faced with because of AIDS. Lung cancer causes 40,000 deaths a year and is the commonest cancer in men and the third most common in women. Regular smokers' chances of getting lung cancer are about forty times those of non-smokers, depending, of course, on how much they smoke and how long they go on smoking.

Smoking also increases your chance of contracting cancer of the mouth, throat, gullet, bladder and pancreas and it is responsible for about one third of all cancer deaths in Britain. It causes death from other diseases, too, such as heart disease, bronchitis and emphysema.

The Royal College of Physicians explains it like this: if you take 1,000 young male adults in England and Wales who smoke cigarettes then, on average, about 6 will be killed or injured on the roads but 250 will be killed before their time by tobacco.

It does seem, though, that the message is beginning to get across to many young men and they either never start smoking or give it up. Fewer people smoke nowadays but it is a sad fact that the only major group among whom smoking is actually increasing is young women. No one is quite sure why that should be – whether it has to do with lack of employment, lack of opportunity or the success of cigarette advertising by the companies who see young women as the only growth area in the market. Possibly for a young girl whose only chance of work is as a shelf-stacker or something similar, the commercial showing the glamorous young thing on a beach or at a party, surrounded by males and looking out into the distance through beautifully made-up eyes as she drags at a cigarette, seems the height of sophistication.

But women who smoke run extra risks. It can reduce fertility, produce smaller babies, bring on an earlier menopause as well as putting them at risk from lung cancer or heart disease. There is no doubt that smoking-related diseases are an enormous burden on the health services.[8]

It now seems that a larger proportion of people smoke in social classes IV and V than in Classes I and II but the same does not apply to alcohol. Alcohol abuse cuts right across the social class barriers and is entrenched in many of the professions, including the medical profession. In fact if you are hard up it is not very easy to booze your way through the week. You can drink yourself to death easier and faster if you have money, whether it is your own money or your expense account. Lunchtime O'Booze is still alive and well and living in Fleet Street, the City, the cosy club and wherever else people meet for a chat and a drink.

At the risk of sounding like a killjoy and as one who enjoys a drink myself, I have to say that there is a pretty dreadful list of diseases that can be caused by excessive drinking. They include hepatitis (inflammation of the liver), cirrhosis (disease of the liver), stomach disorders such as gastritis, bleeding and ulcers, cancer of the mouth, throat and gullet, brain damage, high blood pressure, muscle disease, problems with the nervous system, sexual difficulties, depression and other psychiatric disorders and vitamin deficiency and it can add to the problems suffered by those with diabetes.

Heavy drinkers can be fat and overweight yet actually suffer from malnutrition because they replace food with alcohol and are insufficiently nourished. Women who drink when pregnant pass the alcohol on to the baby across the placenta. In fact alcohol has more adverse effects on women than it does on men. One reason is the water content of the body which, in men, is between 55 and 65 per cent of body weight but in women is only between 45 and 55 per cent. As alcohol is diluted by body fluids, alcohol in men is more diluted than alcohol in women.

Like smoking, alcohol-related diseases also put an unnecessary burden on the NHS but it is not only the diseases that do this. Alcohol is a major cause of road traffic accidents and one in three of the drivers killed on the road have levels of alcohol in their bodies which are over the legal limit. Road accidents after drinking are the single biggest cause of death among young men in Britain.[9]

Diet also contributes to our health. For many people who are living on or below the poverty line there is precious little choice. It is simply not possible to go on to what might be called a health food diet. But, even so, most people in Britain eat far too much fat and sugar.

Ever since Queen Elizabeth I had to smile carefully to avoid showing blackened teeth, rotted by sweetmeats, the British have had a collective sweet tooth and we are the biggest sugar eaters in the industrialised world. Yet all sugar does is add on unnecessary calories and rot our teeth. We love fatty foods, too – chips, fry-ups, cream cakes and so on. They all help us to put on weight, which in itself throws extra strain on the heart as well as pushing up the rate of cholesterol in the blood, affecting both blood pressure and the heart. Heart disease does seem to run in families, too, and recently the British Cardiac Society called for the screening of children from families where a parent has died young of heart disease, so that preventive efforts could be made such as advising a low fat diet, emphasising the need for exercise and the need to avoid starting to smoke.

Just how much diet is responsible for the nation's poor health is difficult to quantify. There are very powerful lobbies against any criticism of the national diet from the food industry and the breweries. The now defunct Health Education Council blamed the Thatcher Government's decision to disband it on extreme pressure from these industries. Certainly, poor diet goes hand-in-hand with ignorance and deprivation although the message does finally seem to be getting across, as witness the growing popularity of wholemeal bread.[10]

It is not possible within the remit of this book to look at drug abuse, although it cannot be emphasised sufficiently that the resources to deal with it are totally inadequate. While there has been a good deal of brave talk by Government spokesmen, including the Prime Minister, about fighting the drug problem, actions speak louder than words and not only has there been little in the way of extra funding (apart from the much-criticised, inadequate advertising campaign), but funds have actually been withdrawn in some areas, a withdrawal necessitating cut-backs in clinics – as has been noted elsewhere.

But, in fact, by far the largest number of drug addicts are perfectly ordinary people hooked on tranquillisers or, to a lesser extent, on anti-depressants. It is estimated that while there are around 50,000 known heroin addicts (although the figure could be much higher) 3.5 million people are at risk from tranquilliser addiction: 40 million prescriptions for them were issued in 1980 alone, at a cost to the NHS of £30 million.

According to TRANX,[11] the National Tranquilliser Advisory Council, 35 per cent of users take them for longer than six months although they are medically useless within a comparatively short period. Three times as many women are prescribed them as men, but the number of men being given tranquillisers and anti-depressants in areas of high unemployment appears, on anecdotal evidence, to be growing.

According to TRANX many doctors still do not believe

that they are addictive and they are increasingly prescribed as an answer to everything, from unemployment and marriage breakdown to bad housing and poverty. This is to use tranquillisers as a barrier between the users and the real world, treating the symptoms not the cause. The drug companies bombard doctors with tons (quite literally) of advertising material purporting to offer panaceas for very real problems such as stress from high-rise housing (Triptagen) and domestic troubles (Prothiaden) to what is merely the common lot such as difficulties with organising housework (Limbritol).

Repeat prescriptions are still easy to obtain and there are plenty of well-documented cases of people being able to ring up and get, say, 100 Valium tablets, without anyone checking up on how many they have been taking and for how long. Yet withdrawal symptoms can last for nine months and be severe. They include stomach cramps, eye defects, breathlessness, loss of use of the legs, dizziness, nausea, panic attacks and mental problems such as agoraphobia, to say nothing of the physical side-effects some of these drugs cause when actually being used. A leaflet put out by the charity MIND makes the point about tranquillisers when it says: 'Remember that anxiety is usually a healthy response to things that threaten your health or well being.'

In view of this it seems to be a particularly senseless decision, at the time of writing, to allow the possibility that TRANX might fold for lack of funds. TRANX was started by a woman who was first given tranquillisers when she was a battered wife. She later became an addict and remained so for seventeen years. TRANX now has a national network of self-help groups trying to make it easier for people to come off these drugs, and the good work they do is not in dispute.

There are, of course, many other ways by which we can collectively damage our health but, to return to where we began, with the Black Report, there is surely overwhelming evidence that the cycle of poverty and deprivation brings with it much of the rest, including ignorance of the need for a proper diet and the dangers of tranquillising drugs.

The polluted environment is a subject which would make a book in itself and it is only possible to mention some aspects of it very briefly here. Enormous and powerful lobbies, including the giant multinationals, are ranged against those who would say that environmental pollution affects the nation's health.

Yet time and again they have been proved wrong. To take only one example, that of asbestos, for nearly forty years the British Government of the day kept to itself the knowledge that asbestos dust caused a range of diseases including a specific form of cancer. It was not until the mid-1970s that the extent of the problem became public knowledge with the publicity given to what had happened in one small town alone, Hebden Bridge in Yorkshire. Here Cape Industries' asbestos mill, Acre Mill, hit the headlines when it was discovered that, between 1939 and 1970, nearly 300 people out of 2,200 workers employed during that period had developed asbestos-related diseases. Yet in the July of 1976 the Asbestos Information Committee, made up of members of the main asbestos companies, launched a £½ million publicity campaign to assure the public that asbestos was safe.[12]

By 1983 107 people had died of asbestos-related diseases in Hebden Bridge, including at least one wife who had merely washed her husband's work clothes. It had taken until the end of the 1970s for the government to ban blue and brown asbestos, and then in 1983 the supposedly 'safe' limits to which asbestos workers could be exposed were suddenly halved. This still 'safe' product was then hurriedly ripped out of schools, hospitals – even the House of Commons. By the time the American actor Steve McQueen had died of the form of lung cancer associated with asbestos and had blamed this on working with the substance when he was a student, there were few takers for the notion that any level of asbestos was 'safe'.

The asbestos story is worth remembering when one hears all those reassurances from chemical firms which spew out pollution into the environment and then tell those living near by that the birth defects of their children and the diseases experienced by both humans and animals living in the area have nothing whatsoever to do with what is being manufactured inside their plants.

Even the water we drink in this country is polluted. By the end of 1986 evidence was emerging that it had been the worse year on record for rivers and streams poisoned by farm waste. The pollution had been caused by the intensive farming practised by so many farmers. Pollution has included slurry, liquid from silage – silage cut and stored through the winter has a high waste liquid content and if it spills into streams and rivers undiluted it destroys the oxygen in the water – and nitrates. At the time of writing, the EEC is threatening legal action over our failure to bring the nitrate levels in our water down to EEC standards. It is known that very high levels can harm babies and are suspected of causing stomach cancer. Yet the chemical firms who press farmers to use nitrates are fighting hard to avoid there being any controls over their use although the massive crop yields they are advertised to promote now end up as Common Market grain mountains.

We now also drench our land with about 1,000 pesticides. Some of those used in Britain are not allowed in other industrialised countries – we seem very brave at allowing substances to be used here which are considered doubtful elsewhere – and some at least were found to have been passed as 'safe' in the USA on the basis of data later found to have been falsified.

At least ninety pesticides cleared for use in Britain have been linked with cancer, birth defects or genetic mutation, yet, according to the London Food Commission,[13] government surveillance of residues in food would be able to detect less than one third of permitted chemicals. In a report published in 1986 it listed 426 pesticides and rodenticides cleared for use in 1985 and noted that 49 were linked with cancer, 31 with birth defects and 61 with genetic mutation. In a recent survey 30 per cent of foods sampled had been found to contain detectable residues. The report criticised the surveillance of residues in food, noting that the most exhaustive testing, carried out by the Ministry of Agriculture, Fisheries and Food, would only identify 110 of 400 permitted residues.

Furthermore, local authority cut-backs had forced public analysts to test for as few as 10 per cent of permitted residues in foods. Dr Tim Lang, Director of the Commission, said that pesticide residues are now 'a potentially major health problem and monitoring and control by the government is totally inadequate'.

In fact the whole area of which pesticides and chemicals are currently allowed and why they should be considered safe is shrouded in secrecy, often in Official Secrecy. Scientists who sit on government quangos such as the Advisory Committee on Pesticides, the Committee on the Safety of Medicines and the Committee on the Medical Aspects of Food Policy (rightly known by its acronym COMA) all have to sign the Official Secrets Act.

If it is not Officially Secret then it appears to be suppressed. The *Guardian* of February 2nd, 1986 carried details of an internal Ministry of Agriculture document known as the 'Register of Environmental Achievements' which appears to be a form of newspeak. It says, for example, that publicity about the Ministry's fight with the Department of the Environment to get British Nuclear Fuels to cut down on radioactive waste discharged from Sellafield is likely to be counter-productive and that publicity about exposure to radioactivity through agriculture and the food chain is 'not desirable'! It goes on to say that while pollution in general should be 'openly discussed', no publicity should be given to the effects of acid rain or to dioxins. (The dioxins are among the most toxic pollutants known to man. One appears during the manufacture of the weedkiller 245T – one of the constituents of the Agent Orange dropped in Vietnam – and another is caused by improper incineration during a chemical process.)

Nor would it be desirable to give publicity to the statutory controls now available to Ministers in the event of the escape of a hazardous substance which might be dangerous to people who had eaten food poisoned by such an accident. There should be no publicity given to the tests on concentrations of industrial chemicals in fish which have been conducted with other EEC countries nor to the work designed to set standards on pesticide residues, or the Ministry's investigation into the effect of a reduction of lead in petrol in food; nor on how wildlife can pass on diseases such as salmonellosis to farm animals.

A series of experiments which were conducted and which appeared to show that cutting down on the amount of pesticides used by farmers appeared to have little effect on the subsequent crop yield and quality of the crops grown was also suppressed. Instead the document said the Ministry should give publicity to a 'rational' use of pesticides. Taken as a whole, says Richard Norton-Taylor, who wrote the story, the document 'reflects the extreme sensitivity – sometimes bordering on schizophrenia – in the ministry about discussing environmental problems'.

At the time of writing the DHSS is also sitting on a report into the apparent 'clusters' of leukaemias and cancers around nuclear plants, but that is another story.

The problem of environmental pollution is that it is very hard to prove that any particular case, or series of cases, has actually been caused by a particular type of pollution – not least because very often the effects are long term. For example, cancer can appear anything up to twenty years after someone has been in contact with a carcinogen.

So it is just not possible to say how many of our current health problems are caused in this way. But it is not an issue that is likely to go away. Usually, by the time we do know that there has been a hazard, it is too late for some. Tombstones cover all too many of those who were told that asbestos could not harm them.

6

'I Wouldn't Start from Here . . .'

When one looks at the state of the National Health Service today one wonders what can be done about it. Those contemplating how it might develop over the next few years might well feel like the city slicker in the old joke who asked a rural yokel how to get to the nearest town and was told 'If I were you I wouldn't start from here!'

The estimated cost of the NHS in 1985/6 was £17.5 billion, yet patently it has been insufficient to meet all needs. In part, as we have already seen, this is because inflation within it has risen faster than inflation nationally; and that is due to a number of reasons, including new technologies and the fact that some of the increased funding has gone on salaries and administration.

Yet opinion polls over the years show a steady majority in favour of more spending on the NHS even at the expense of tax cuts; indeed a majority feel that not enough is being spent on it, that majority varying over the last two years between 67 per cent and 87 per cent depending on which poll you look at.[1] The Institute of Health Services' management report, quoted earlier, says: 'The polls present a picture of increased public concern with levels of spending on the NHS, even when it is clearly understood that the consequences of greater spending would be higher taxation.'

However, it is also fairly obvious that more money is not the only answer. To the outsider the NHS looks like either a great lumbering dinosaur or a bulging wall holding back dammed-up waters, where as soon as you plug one leak another appears somewhere else. The NHS was already littered with tiers of management, committees, departments and so on before the new incoming managers arrived and brought in yet more staff. One District Health Authority official complained recently of the sheer volume of paper pushed around between departments. 'The person who invented the photocopier has a lot to answer for,' he said, gloomily. As in any huge businesses, plenty of empire building has gone on, so there would be in-built resistance to truly radical change.

None of the political Parties, on present showing, seems prepared to do more than tinker with the bureaucracy, obviously baulking at the notion of taking on so many entrenched interests. Regional and District Health Authority members being appointed, not elected, tend to follow whichever party line is their own.

During the years of broad consensus on the NHS, the Conservative Governments devoted most of their energies to reorganisations within the existing structure. Now the emphasis is on hiving off bits and pieces to the private sector, encouraging private medical insurance and trying to run it as a business. Money, the present Government continually tells us, is not all, which is correct; but proper resourcing is at least part of the solution. All the indications are that with a third Conservative Government returned, we shall be in for more of the same, with money either taken away from one part of the service and suddenly reallocated to another as 'new money' or extra tranches of funding found every time a General Election looms. However, in one respect at least the present Government is to be commended, and that is for its wide consultation on the direction in which we should be going with regard to primary care – the non-hospital medical services. We shall return to this in the next chapter.

Those looking for a radical new look from Labour will be in for a disappointment as well, if what they propose in their latest document 'The Best of Health' is the manifesto for the years of Labour government. It has plenty of sensible suggestions to make, most especially, again, in the area of primary care; but it shies away from tackling the bureaucracy. However it does lay a good deal of emphasis on preventive medicine and health education – hopefully it might even reconstitute the Health Education Council, which the Thatcher Government has just abolished.

The Alliance, too, laid more emphasis on preventive medicine and health education than the present Government. But all three parties still see the National Health Service as a *Health* Service – not the Sickness Service which it has manifestly become, as suggested in an earlier chapter.

The aim of this book is not to come up with solutions but to encourage debate on where we go from here, and also to look at some new ideas and at one past example of how health care in the future might develop. Among the major changes which have taken place since the NHS was set up has been the rapid growth of the consumer movement within it and it is surely no longer possible to go forward without far more consumer input.

The National Health Service was set up in such a way that it has an in-built patriarchal structure. This was almost inevitable. It was devised by politicians who decided what was best for people, run by civil servants trained to carry out the orders they were given, funded by the Treasury to whom all Health Ministers have to go every year, cap in hand, when the amount of public expenditure has to be decided. At senior level, naturally, it is staffed by doctors who, with a handful of rare exceptions, tend to be among the most conservative of professionals, possibly only lawyers being more rigidly set in their established views.

So from the beginning – and until relatively recently – it has all been on a 'them and us' basis. 'They', whether government or the medical profession, have known what was best for us. Vital decisions have been taken in secret – in some cases in Official Secrecy – with token consultation only, and mainly with those bodies to which selected members of the public are appointed by the very Health Authorities they are supposed to monitor.

Whatever route we travel by in the future we will need a service to look after us when we are sick; but that is only half of what a true National Health Service should be about. That half will require adequate funding, although this may well be easier if the second half of the plan is carried out – *positive* action on health which would mean less pressure on the sickness side of the NHS.

If we do not begin to think along these lines then the NHS as it is today can only get worse and worse. What will happen then will take us a long way away from the ideals of 1948. Certainly many more people will opt for private health insurance but, as the system operating in the USA shows, that by no means ensures that you are all right, Jack. Private medical insurance companies do not want to take on the chronically sick, the old, those with problems which will obviously get worse. If you persuaded the better-off to take up private medical care, anyway, you would still have to fund some kind of basic system for the poor, which would lead to a two-tier system of health care. In America there are those so poor that they do not even qualify for the basic Medicare plan and are literally without any cover at all apart from a handful of charity hospitals.

If we battle on with slightly more funding but the philosophy still much as it is, then there is no doubt that the rationing of services, which is now happening covertly, will have to operate quite openly. There are those in the present Government and among their advisers who say outright that expensive treatments should be rationed and carefully examined to see how cost-effective they are. Already doctors know that they have to turn away deserving cases through lack of beds or operating time or the necessary facilities, have to weigh up which out of two or three people should get dialysis treatment or a kidney transplant, who should have the really expensive heart or heart-lung transplants.

A system has now been devised to try and measure who should have what using a unit called a QUALY. QUALY stands for 'Quality Adjusted Life Years', a phrase which led at least one commentator to remark that it had all the inherent charm of the Final Solution . . . Much of the work on QUALY has been done by Professor Alan Maynard, Director of the Centre for Health Economics at York University.

To work out a QUALY you not only need to know what a treatment will cost, but also need to have some way of evaluating what its outcome is likely to be. Outcome, in this sense, means that you have to find some way of measuring the increase in length and quality of a patient's life. An example might be that of a hospital department having a budget of £20,000. You might generate one QUALY by spending the whole lot on renal dialysis. Or you could spend it on replacing arthritic hips and generate twenty-five QUALYs. So one procedure offers twenty-five years of full quality life, the other only one year for the same budget.

Another example would be a seventy-year-old arthritic person with a hip which needs an artificial replacement, one of those qualifying for one of the twenty-five QUALYs on offer above, you might think. The doctor would then sit down at his computer, into which had been fed all the medical and other information known about that person, and find that such a hip operation might give this particular person an additional ten years of high quality life, ten QUALYs at a cost benefit ratio of £200 per QUALY. This seems to be quite a good return on investment, so if it is possible to get him into hospital in the near future then he or she can have his/her hip operation. However if he or she were older, or even younger, but had some other condition which might cause future deterioration, then the amount of 'full quality life' left might be quite small – say half a QUALY – and the cost of treating him would not be considered affordable by the NHS.

Other examples come to mind, such as trying to work out whether someone suffering from cancer has enough QUALYs to be deserving of any other than minimal treatment or whether it is 'worth' saving sick and/or premature newborn babies. In an edition of the television programme *This Week* transmitted in January 1987 Professor Maynard expanded on his theme of QUALYs to some pretty horrified consultants, some of whom suggested he might try explaining it all to their patients. The paediatrician who had a special baby care unit was particularly appalled and spelled out to the professor that his view that saving such tiny babies came pretty low down the QUALY list – for 'if sacrifices have to be made few people grieve over a tiny baby' – did not even make good economic sense, for in many cases the baby would not actually die but, without treatment, would become permanently handicapped and thus be a charge on the State and the NHS for many years.

Now a third Conservative Government has been returned we may hear a lot more of Professor Maynard's QUALYs.

So that is one way of trying to deal with the problem – by rationing. One can only feel very real sympathy for the doctors who would have to operate such a system all the time, who lived and who died, who was treated sufficiently well to be able to lead a good life and who had the bare minimum for survival.

But let us assume that we do not have a rationed hospital service. Then, if we were looking for an ideal system, we would still need to think how it might be made to work in a way which would better benefit both staff and patients.

This would not only mean replacing out-of-date buildings and equipment and ensuring adequate staffing, but making sure that the hospitals, wards and medical staff are in the right places. This might seem obvious, but decisions taken centrally in Whitehall and then passed on to a Regional Health Authority office which is, itself, many miles from the district areas which it is supposed to serve, can mean that even a new hospital or an improved ward can be built or opened in the wrong place. In a rural area, for instance, too much centralisation of services – however good they are – can mean long and worrying distances for those who use the services and those who visit them in hospital especially if they have to rely on public transport.

The trendy word 'overview' is pretty horrid as a word, but it does describe what is needed when consideration is given to physical buildings. It is not necessarily cost-effective for patients to have to be trundled in miles and miles by ambulance nor for their relatives to have to find money for taxis to get them to the hospital because deregulation of the bus services means that they no longer have any buses at all – this is already the case in parts of south west England. It is also disgraceful in this day and age that some health authority areas are without what has now become basic equipment in most European countries, items such as body scanners and specialised equipment for treating cancer. All praise to those bodies who raise the money charitably for special items; but they should not have to be relied on to do so. (There have been cases, too, when only one item, such as a brain scanner, exists; it breaks down, and the patient has to be taken many miles away to another hospital which has one in working order. Sometimes by the time the patient arrives it is too late.)[2]

But how and where hospitals are to be built and how they should be equipped is not within the remit of this book. Obviously there needs to be far wider consultation than hitherto and this raises very real problems about how you set about doing this within the existing bureaucracy of the NHS without yet another costly and upsetting round of reorganisation.

In fact a whole range of bodies has already had a substantial effect on hospitals and the way they are run, none of which existed when the NHS was first set up. As well as organisations for specific groups, such as the Spastics Society and those for the mentally ill and mentally handicapped like MIND and MENCAP, there have been the many pressure groups which have brought about changes which have since become established practice.

It is completely accepted now, for instance, that mothers should accompany their small children into hospital, whenever possible, and have ample access to them while they are there. Yet this was almost entirely due to the efforts of NAWCH, the National Association for the Welfare of Children in Hospital. Prior to the emergence of NAWCH, for years young children were left behind screaming while their mothers were told they would settle down much better if they did not see much of them.

It was maternity pressure groups like AIMS, the Association for Improvements in Maternity Services, who brought about the now accepted practice of fathers being able to stay with their wives throughout the delivery of their babies if they so wished as well as pressing for more 'humanity in obstetrics', though with the advent of 'high-tech birth' this is a battle which is still going on.

The Patients' Association grew out of the overwhelming need of patients themselves to have a say in the system in general and how they were treated in particular. The NHS grew up in the then prevailing atmosphere that the patient was some kind of a hindrance to the smooth running of the hospital and its wards. Everyone could get on so much better without them. As well as being a patriarchal system the NHS is a tremendously hierarchical one, too, with the consultant at the top of the medical tree.

Anyone treated in hospital some years ago – and in some places still today – can remember the mounting excitement of the ward round. Patients would be made tightly into their beds even if they had just been given diuretic medicines to make them want to use the lavatory, sheets would be turned down exactly the regulation width, the tension would mount as various outrunners from the Great Man's entourage appeared ahead of him and, finally, the door

would swing wide open and in would come the man himself, complete with registrar, houseman, students, sister,
nurses and anyone else who happened to be around. They
would then stand around the unfortunate patient and discuss him or her in abstruse terms as if he were somewhere
else.

When this happened to me some years ago the consultant
was a person I knew slightly socially. 'Good God,' I told
him when he arrived at my bed, 'at the very least I expected
you to arrive carried in on a litter surrounded by Nubian
slaves with fans.' The nursing staff turned pale. Some
months later we found ourselves at the same dinner party.
'Go on,' he said, beaming, 'tell them about my ward
round!'

The Patients' Association has not only pushed for recog-
nition of the serious needs and rights of patients, it has
consistently pushed for more and better information to be
given to them.

Much of the pressure from groups like these – and many
others – has borne fruit. There are hospitals now where
doctors and nurses pride themselves on the way they
communicate with patients even when, as is currently the
case, they are under-resourced and overstretched.

One recent example of good practice in communication[3]
is that of a consultant in Northallerton who provides each
patient with printed information guidelines on details of
his surgical condition, what the operation is to achieve,
how it is to be done and what the patient can expect before
and after it, all couched in jargon-free prose. It goes further
and tells the patient how to look after himself at home
when he leaves hospital, when he can drive the car, return
to work, how to get certificates, all the problems often
overlooked in a busy hospital. There are guidelines for
each type of surgery and they are available for the consultant's colleagues too. So far this particular consultant knows
of no similar system in this country and says patients and
relatives have been delighted with his idea. Not least, when
he sees the patient, his time in outpatients' can now be

better spent discussing residual queries rather than spending ages describing the routine problems covered in the guidelines. This is manifestly the kind of procedure which could be copied elsewhere.

This is the new outlook on hospital care; but there are unhappily still areas where the bad old ways still obtain, such as the outpatient clinics where consultants still over-book or book in several people at the same time so that their, the consultant's, time won't be wasted, irrespective of how much patient's time is wasted instead, as instanced in Chapter 1.

Government had recognised the need for an official channel for the NHS consumer apart from the wide variety of pressure groups and health charities and other bodies, and in 1974 set up the Community Health Councils. Each district has its own CHC representing local NHS users. Its membership is drawn from Regional Health Authority appointees, members of local authorities and representatives of voluntary organisations, none of whom are elected. The CHC is serviced by a permanent secretary and, when it can be afforded, clerical staff.

The work undertaken by a CHC is as good as its membership, a truism but none the less valid for that. Its main activities include:

– advising the NHS on users' views of local services which, with regard to hospitals, would include things like waiting time in outpatients' clinics, standards of facilities and care;

– advising its Regional and District Health Authorities on plans and developments, in particular the closure of wards and hospitals;

– giving advice and information on health services to the public;

– assisting people who want to make a complaint up to any level;

– promoting the interest and involvement of local people in their own health and in the NHS.

With regard to the problems facing the hospital service, the CHCs have undertaken a massive amount of work, nationwide, surveying hospital facilities within their own areas. As things are at the moment their surveys and reports are a vast underused resource for those planning future services.

If there is really a will to make the best use of hospital services, then one way of assuring direct consumer input would be to take on board the results of CHC surveys, which represent the views of those who actually use every part of the hospital service.

Community Health Councils, most of which are linked together within the Association of Community Health Councils, work across all political Party lines. In 1985 the Conservative MP Michael McNair-Wilson found himself in an NHS hospital as an emergency admission. After he had recovered somewhat from a serious illness, he began to formulate the idea of a 'Patients' Charter' which would set out the rights of those who are ill.

After consultation with him and with various bodies, the Association of Community Health Councils produced such a Charter, which has been presented to all the political Parties. It appears in full as an appendix at the end of this book. All persons, it says, have a right to health services appropriate to their needs, regardless of financial means or where they live and without delay; to be treated with reasonable skill, care and consideration; to be informed about all aspects of their conditions and proposed care unless they express wishes to the contrary; to accept or refuse treatment; to a second opinion; to be discharged from hospital only when adequate arrangements have been made for continuing care; to privacy for all consultations; to be treated at all times with respect; to make a complaint and have it investigated thoroughly, speedily and impartially. There is also a good deal more.

But perhaps this is the place to look at one particular issue: what happens when something goes wrong. There have been a number of well-publicised cases where accidents have happened, including that of David Wood-house, a young SAS man who went into hospital in 1981 for a routine appendix operation and never came out of a coma.

The Health Authority immediately took all the necessary steps to set up an independent enquiry into what went wrong but this was completely negated by the Medical Defence Union, who refused to allow any of its members to give evidence. This is an all too normal course of events because of the British system of dealing with matters of this kind, the adversarial system.

At present the victim or his/her family has to prove negligence and take the health authority to court. It can take years for the case to be heard, and can be extremely difficult to prove, for the 'victim' is often denied access to the evidence; it is usually financially prohibitive as well, with Legal Aid difficult to get, and very often the injured person just gives up in despair.

The Association of Community Health Councils for England and Wales (ACHCEW), among other bodies, is now pressing for what is known as a 'No Fault Compensation Scheme'. Sweden and New Zealand already have versions of this working successfully. The suggested name for it here would be the 'Patient Insurance Scheme'. Under such a scheme a patient could be offered compensation very quickly and then an enquiry could be held into what went wrong without the patient having to prove negligence.

The Swedish scheme seems to be the better of the two. In fact there are three types in Sweden, one dealing with industrial injuries, one with the adverse effects of medical drugs and the other with medical accidents. The scheme is administered through a consortium of insurance companies with premiums paid through the local authorities, who administer Sweden's health service.

It offers fast compensation for medical accidents – over 50 per cent of claims are settled within three months. It is not necessary to establish negligence on anyone's part for compensation to be paid. Claims are initially decided by the insurance consortium, but if their decision is not acceptable to the patient the matter is referred to an independent Patient Injuries Committee. If the patient is still not satisfied he/she can go to arbitration and, as a last resort, going through the system does not preclude the patient from taking legal action. Few people now find this latter option necessary.

The benefits should be obvious. There are no costs incurred by the patient, no financial gamble is involved. It encourages more openness by the professions. If there is a doubt the benefit of it can be given to the patient without adversely affecting the reputation of an authority or staff member; it is quick in most cases and, finally, periodic reviews can identify unsatisfactory outcomes of currently accepted good practices and encourage research into ways of improving accepted standards rather than merely concentrating on deviations below them.

7

The Sharp End

We have come a long way from the crumbling hospitals of the beginning of the last chapter, rationed health care and the assumption that there is very little that can be done apart from throwing money at it all.

But this is where the next part of the National Health Service – the health part – should come in, to ensure that as few people as possible actually end up needing the acute care provided by the hospital service. This brings us to Primary Care: i.e. treating people in the community, preventive medicine and health education.

It is obvious from the foregoing that most of the pressure for better services has come from the informed, articulate, professional middle-class, who can do their best to ensure that they receive adequate treatment as well as pushing for better services for all.

However, if we want better national health we need to target services and funding to those groups who are so much more at risk, the poor and the deprived.

Whatever may be the differences of opinion as to the problems involved and how they should be tackled, there is now virtual unanimity that resources and effort must be put into the primary care services (that is, care not involving hospital treatment).

From cradle to grave there are none of us who do not use these services at one time or another. We visit our local GP. This experience can vary a great deal, from the country surgery and the dwindling band of doctors who still behave like those in the old TV serial *Dr Finlay's Casebook*, to the modern purpose-built medical centre with its practice manager and forward-looking thrusting doctors, and everything you can imagine in between.

The experience of visiting the doctor can vary. You can be given an appointment and seen roughly on time, the receptionist, either on the telephone or when you arrive, can be helpful, the doctor may listen to what you say and not try to hurry you out of the room or begin writing a prescription before you have finished your opening remarks. Or you can arrive even with an appointment and wait and wait . . . that is, if you are able to get through the receptionist and get an appointment. The Dragon Queen who guards the doctor and makes decisions for him as to whether you are sick enough to be seen – let alone that he should actually have to come out to you – is still alive and well and living in far too many GP surgeries throughout the land.

If you are new to an area then you have to cast around for a doctor. How do you find out who is the best or most suitable one for you, supposing there is a choice? Or what services that particular practice offers? Or if any of the doctors in a partnership have particular qualifications and skills? The answer to all those questions at the moment is 'with great difficulty'.

What happens if you are unhappy with your doctor and want to go to someone else? It can be very difficult to move and people are very often frightened to do so even if there is a choice. In some rural areas there is no choice. If you do move elsewhere do you have the label 'trouble-maker' hung around your neck?

What happens if the worst comes to the worst and you have to make a complaint? We shall look at that more closely later but that, too, is made very difficult. Even if your complaint is upheld at the level of the Family Practitioner Committee and your doctor has already been disciplined before for the same kind of thing, you won't be able to find out because it is all kept secret.

Perhaps you want a dentist; how do you find a good one, or any dentist, come to that? GPs, while being private contractors do, if they are contracted to the NHS, have to treat patients on the NHS. Dentists are in no such position. If they want to then they can treat only private patients, a fact some unfortunate people only discover when they have arrived in pain on the surgery doorstep.

Your pharmacist when he hands over your NHS medicine does not have to do more than grunt 'three times a day it's on the bottle'; and you more or less take what your optician tells you on trust.

For many, particularly the elderly, the housebound or those with young children, primary care means the community nurse, the health visitor or the midwife. It means the physiotherapists and speech therapists who work outside the hospitals, the chiropodists and that whole range of services which is grouped under the general heading of 'primary care'.

That is primary care seen from the basic point of view of the family, especially the family with its breadwinner in work, with no chronic sickness and with no elderly or handicapped relative to be looked after at home.

But the demand on the primary care services goes much further than that. It is a startling fact that there are now more women staying at home to look after elderly and handicapped relatives than there are women looking after babies and small children. This is partly due to the rapidly ageing population already mentioned and to the move to 'care in the community'.

Of all the NHS hospital services, those for the mentally ill and mentally handicapped were the most run-down and understaffed. Over the years we have been fed with a diet of horror stories about life inside some of the large institutions, housed in buildings designed as the original 'lunatic asylums'. It has been a vicious circle of lack of resources – under all Governments – leading to lack of staff, lack of care, untrained and inexperienced workers and poor or almost non-existent facilities. While some of the old mental hospitals, however under-resourced and hard pressed, have managed to provide standards of care remarkable in such circumstances, others manifestly have not.

There will always be those who, for a variety of reasons, can never be catered for outside a closed and/or hospital-type system. But for the rest, the return to the real world outside seems the most humane notion, along with all the necessary back-up services. The crunch is in the latter phrase. For it is on families that the burden falls and now it is not only the large mental hospitals which are being closed down before adequate alternative arrangements have been made, but even small custom-built mental handicap hospitals already securely based within local communities. In these cases, for 'community care' read cutbacks in NHS funding.

For community care is not cheap. Those looking after the mentally and physically handicapped need all the help they can get. Yet local authorities themselves are being forced to cut back. The carers may need home help, day centres for those they are caring for, respite care when they need a real break, the regular services of a physiotherapist and enough money to live on without eternally worrying about paying the heating bills or the rates. Care of the mentally and physically handicapped involves all the primary care services, particularly the doctors and community nurses.

The same is true of the elderly. Since 1979 pressure has been put both on local authorities and on the NHS to leave the care of the elderly to the private sector when they cannot be looked after at home, and this has led to a mushrooming of private residential homes and nursing homes for the elderly. Boarding house keepers, especially in holiday areas, have been quick to cash in on the generous terms offered by the Government. Owners of 'Sea View' and 'Beachside', watching British holidaymakers flock away to the Costa Packet where sun and cheap booze is assured, have turned their bed-and-breakfast villas into the 'Sea View Home for the Elderly' or the 'Beachside Nursing Home'.

There are inherent dangers in this. First, if the holiday trade picks up then there is nothing to stop all those ex-boarding house keepers turning their residential homes back into boarding houses and turfing the old people out literally on to the street for the local authority or NHS to deal with. Second, even now, while those running such homes are often happy and willing to do so as long as their charges are reasonably active and fit, it is a different story when they become frail, inactive, have failing sight or become incontinent. Then, unless they can pay for real nursing care as distinct from ordinary staff, the search is on for an NHS hospital bed to get the old person back into the State system.

Lastly, while some homes are of the highest standard, a proportion are not, as we are often reminded in the media. Apart from being in totally unsuitable situations, such as at the top of steep hills or in the middle of fields as with some homes registered in the early days, the conditions for residents can be grim, with a lack of basic facilities, insufficient linen, poor and ill-prepared food (or even, in some cases, insufficient food), a lack of staff and a general air of neglect as the owners seek to make as much money as they can out of the system. Standards are checked by only a handful of inspectors, who will be lucky if they can get around to a residential home once every two years. Community Health Council members who can visit hospitals at any time to monitor standards of care and conditions are not allowed into the new private residential homes for the very reason that they *are* private.

Nor do the problems stop there. An influx of private residential homes into a small town throws a large additional burden on to the GPs, who can suddenly find themselves with a large number of extra elderly people needing their attention and have to take them on whether they want to or not.

So far we have looked at two major sections of the community that require primary care – the ordinary everyday person in the street and those requiring 'care in the community'. There is a third, the poor and deprived to whom we have already referred.

Two further reports have come out recently which back up the findings of the Black Report. They appeared in February 1987, within a day of each other. 'Heartbeat Wales'[1] was a government-sponsored survey of 22,000 people living in that country. It found that without a shadow of doubt unemployment brings ill health and that both the unemployed and low-wage manual workers were likely to suffer more from angina, respiratory illnesses and raised blood pressure than high earners.

Within the manual groups 32 per cent are more likely to die of heart attacks and strokes, and the health gap is still widening. Smoking is partly to blame for the health differences between social groups I and V with the unemployed smoking the most of all. Poor diet also plays its part, along with drink, more unemployed men drinking to excess. Sadly, those in the manual and unemployed groups felt they had less control over their ability to resist ill health than those in higher income groups.

The second report was the basis of a *This Week* documentary transmitted on February 19th, 1986. It featured a health survey carried out in Sheffield, street by street, ward by ward and district by district into death and illness. It proved that from cradle to grave your chances of dying before your time were twice as high if you lived in a poor area. Children from poorer areas were three times as likely to be admitted to hospital with chest problems and five times as likely to be admitted with stomach problems as their counterparts who were better off. The incidence of heart disease, respiratory problems and cancer doubled, or even trebled, between different wards and different districts. You were marked out by your class. The conclusion of the film was that primary care resources, as has already become apparent, must be targeted particularly on this large group.

Yet how can we possibly cope with so many and varied demands on the primary care system?

There are many suggestions and many different views. In 1986 the Government produced a major Green Paper on primary care. This followed on from the Report of the Cumberlege Committee, under the chairmanship of Mrs Julia Cumberlege, on the future of the nursing service and on a document on the complaints procedure. To give credit where it is due, the Social Services Secretary, Norman Fowler and his Department embarked on one of the most extensive consultation procedures we have seen.

To ensure that the widest number of interested bodies could participate, consultation was in the form of a 'road show'; those taking evidence set up their camps in different parts of the country where those who used the services could give their views. It could only be a welcome move that so much emphasis was put on the views of the users of the services, not just on those of the providers.

The aim of the government Green Paper, it was said, was to give the consumer both more input and more choice: choice in obtaining high quality primary health care services. It also encouraged the providers to aim for the highest standards, sought to provide the taxpayer with the best value for money from NHS expenditure on the family practitioner services and hoped to set clearer priorities for the family practitioner services in relation to the rest of the NHS.

Among ideas put forward were special bonus payments for good practice by doctors and American-style 'Health Shops'. Health shops would integrate primary care services under one roof and would allow non-professionals to run them as small business concerns. The Green Paper recognised that consumers should have more information on services provided by doctors, but overall it was a disappointing document. Comments on funding, as the Association of Community Health Services said in its response to the document, seem to be confined mainly to the problems of controlling costs. 'Improvements in services are to be affected by the operation of mechanisms analogous to those in the market place.' Services apparently should be 'demand led'. It also appeared that extra resources would have to be taken from the already stretched hospital services.

There were no plans in the document for changing the status of GPs, by making them salaried employees of the NHS rather than private contractors, nor did there appear to be any concern over the rising costs of dentistry.

The Green Paper in the main looked at how services might be provided and funded with a greater emphasis on market forces. Most worrying of all, though, it looked at the primary care services only as they affect the first group of those who use them: the average, reasonably healthy family. There was virtually no provision for extra resourcing for care in the community for either the mentally or the physically handicapped or the elderly, and there was certainly no real emphasis on the needs of the poor and deprived.

The Labour Party, too, wants to increase patient choice. In its document, 'The Best of Health', it recognises the need for greater resources, and sees the need to integrate not only the services provided by the NHS and local authorities, but to look at how, say, public transport fits in as well.

It suggests having salaried family doctors in difficult areas, such as parts of the inner cities, which would offer advantages both in recruitment and in planning a higher quality service. GPs should have smaller lists and work as part of a team which includes nurses, health workers, health education workers and ancillary staff. Doctors should have to produce a Practice Statement giving full details of all the services they provide. Patients would be encouraged to get involved in the work of the local health care team. (The Government posited the idea of 'patient liaison groups' to work in a similar capacity.)

The work of the Family Practitioner Committees would be taken over by health districts which would monitor health needs in their area and, where necessary, liaise with other authorities. It would be they who gave doctors the option of being salaried and provided top-quality deputising services where necessary.

There is mention in the document of both preventive medicine and consumer input, although this would appear to mean a mass of ill-defined committees, some elected and some not. But it does at least recognise that primary health care is not just about organising the providers of medical care.

The Green Paper and ancillary documents were then considered by the all-Party House of Commons Social Services Committee. It made sixty-two recommendations, most of the anodyne type – roughly equivalent to saying that one is against sin. Picking out a handful which are more useful than the rest, it does recommend that the government should bring together the conclusions of *all* its reviews of primary health care, nursing, etc., and consider the compatibility of its conclusions before reaching decisions on primary health care. It recommends the further development of multi-disciplinary services extending beyond the boundaries of the NHS.

It sees no benefit in Family Practitioner Committees being separate authorities (a move made by the present Government) and thinks that there should be negotiations with doctors about alternative ways of remunerating them. It agrees that they should provide more information for patients.

With an ageing population and 'care in the community', plus earlier discharge from hospital, doctors should have smaller lists; this means employing more doctors. The Green Paper toys with the notion of some doctors in inner cities being on salary. It backs nurse practitioners and a better training and career structure for nurses and makes one important point: 'good primary health care teams will be a reality only if doctors and nurses learn to work together'. It expresses concern about the uneven spread of dental services and the spiralling charges. It sees no priority for setting up health care shops. It sees the role of the consumer, whether represented by community health councils or other groups, as being primarily concerned with disseminating information. It doubts that any major new legislation is called for.

Preventive measures, health education and real consumer input hardly get a look in and as for targeting the deprived sectors of the community – well, that is not considered at all.

The immediate response of the doctors' organisations to the Green Paper was fairly negative. They did not like the idea of bonuses for good practice, but they have a point in this, in that it would be extremely difficult to decide, in cash terms, what constitutes good practice. They seemed happy to stay as private contractors and gave a guarded response to the question of providing more information. At the time of writing there seems to be little in the way of new thinking about primary care.

The nurses had already been considering change and, as was mentioned earlier, one of the documents under consideration was the Cumberlege Report.[2] The Cumberlege Report saw a far more active and independent role for community nurses. It recommended that within its boundaries each District Health Authority should identify neighbourhoods for the purposes of planning, organising and providing nursing and related primary care services; that a neighbourhood nursing service (NNS) should be set up in each neighbourhood; that all community nurses in whatever speciality should ensure that their specialist contributions were fully co-ordinated with the NNS; that all other specialist nurses working outside hospitals should be actually based in the community and assigned to specific neighbourhood services, with a commitment given to ensuring that they would have adequate time to do the job properly; that nurse practitioners should be introduced into the primary health care system; that the DHSS should agree a limited list of items and simple agents which nurses could prescribe; and that doctors and nurses should work together as equal partners in a properly integrated primary care team. The rest of the recommendations were mainly administrative.

The Cumberlege Report is a sensible document. But some doctors still do not like the idea of nurse practitioners, seeing them as an encroachment on their own preserves. Yet when nurse practitioners were tried in Birmingham they were not only successful but in general they appeared to reach those very groups that need to be targeted.

In 1982, a general practice in Birmingham employed two nurse practitioners. It had 4,728 patients on the list of which 49 per cent were female. The nurse practitioner offered open access to patients for 15 to 20 minute appointments. She undertook preventive work, health education, the detection of serious abnormalities, the management of chronic and common acute problems ('flu, etc.), limited prescribing from a list agreed with the GPs, reference to consultants and blood tests.[3]

It was interesting that of those who chose to see the nurse practitioner, 72 per cent were women and 66 per cent below the age of forty-nine. Of those seen, 34 per cent were referred to the doctors and 44 per cent made further appointments with the nurse; 60 per cent of the consultations were for preventive medicine, health education, family planning and social problems.

Over the three years the scheme was in operation an excellent relationship was built up between patients and nurses. All those who saw the nurse said they would be happy to return and that they trusted her to refer if necessary. They said that talking to the nurse was 'less embarrassing', 'better for advice' and that 'she will give

you more time'. 'Nurses', they said, 'do it better.' The
nurse was prepared to spend time explaining why a procedure was necessary and also 'listened – the doctor doesn't
have time . . .' 88 per cent of patients thought nurse practitioners were a good idea and more friendly.

The nurses themselves found hospital consultants not
too keen, although they would not explain why. But they
found that they did reach those very groups that are at
risk, according to the Black and other Reports. As well as
offering the range of services already mentioned, the nurse
practitioner also undertook referrals to the social services
department and, when necessary, rehabilitation. She had,
said one nurse who took part, 'a truly holistic approach'.

Early in 1987 the Association of Radical Midwives prepared a paper of their own under the title 'The Vision'.
Midwives have become increasingly dissatisfied with the
way their role has been downgraded as deliveries have
moved almost entirely into hospitals accompanied by what
is known as 'high tech' birth. Those of us who had children
at home under the care of the district midwife can remember how comforting and pleasant it was to be looked
after by one person throughout pregnancy, have the baby
delivered by that person and then continue to see the same
nurse as a health visitor. She became a family friend.

The Association of Radical Midwives want to see midwives return to playing a key role in primary health care,
not just for all those middle-class mothers who are the
most articulate about having their babies at home if there
are no complications, but also so that they can reach the
at-risk deprived groups who often do not bother with
sufficient antenatal care.

The mother, says the ARM, should be the central person
in the process of care and should be able to make an
informed choice over how she delivers her baby with the
full utilisation of the midwife's skills. Childbearing women
should have continuing care, which must be community
based, by doctors, nurses and health staff who are accountable to those who receive it.

Midwives will work either as members of a hospital team who will look after a woman before, during and after the delivery of her baby – along with the consultant when necessary but at all times ensuring continuity of care – or in the community in a community practice. Ideally the ARM sees practices of two to five midwives, depending on local needs. They would not wear uniform and would be known in the community through visits to schools as well as to child-bearing women. They would have access to consultants, to laboratories for tests and if it was necessary to transfer a woman to consultant care this would be with the agreement of the consultant, the midwife and, not least, the woman. Midwives would either go to a hospital to deliver a woman there or, if she wanted it and circumstances allowed, deliver the woman at home.

GPs would provide medical care when necessary; but much of the obstetric care currently given by GPs results in either duplication or failure to make use of midwives' skills, says the ARM. The midwives point to the system in Holland where most babies are delivered at home and where the death rate of newborn babies is substantially lower than it is here. In Holland a woman can choose to book either with a midwife, or with a GP specialising in obstetrics or with a consultant. The majority opt for the midwife.

All kinds of bodies took the opportunity to respond to the Green Paper, including the National Consumer Council and the Association of Community Health Councils and individual local councils. These groups and, indeed most of those who gave evidence, called first and foremost for adequate funding. The Green Paper's preoccupation with controlling costs and market forces appears to militate against a radical rethink about what needs to be provided and to whom.

In the area of basic provision of primary care, both the NCC and ACHCEW, from their own surveys, found that on the whole people were fairly satisfied with the services offered, although there were obvious differences between areas where there were truly adequate services and areas where they were not.

But it does appear, as we have already shown, that there are very real difficulties in some areas, and certainly among some groups, put in the way of those wanting to register for medical care. Elderly people can find it hard to find a GP who suits them even if they do not live in an inner city area or, for instance, a surgery within walking distance. In some areas surgeries are actually being closed down and in Gwent patients were told to look for a new doctor as a surgery was going to close, only to be told by other doctors in their area that their lists were full and there was nothing they could do about it. Patients complained of a shortage of women GPs, long waits for an appointment even if the patient felt the matter was urgent, insufficient time given to that patient when he/she finally got into the surgery, the gradual reduction of surgery hours and the apparent growing reluctance of GPs to make home visits.

There is certainly insufficient information available to allow people, in those areas where it is a possibility, to make an informed choice about which practice they should join. Many doctors have been surprisingly resistant to providing this, although a growing number do now produce practice leaflets. The Scottish Consumer Council approached doctors in Edinburgh about the possibility of publishing a directory of doctors. It met with fierce professional opposition. However the SCC went ahead, listing all the information they could acquire on twenty-one practices. A follow-up survey of consumers showed without a shadow of a doubt that they really appreciated such a service. In fact they suggested to the SCC over thirty other items they thought should be included in it, such as the number of patients on each GP's list, arrangements for children in the waiting room, areas of specific interests of the GPs, more information about maternity care and services for children.

It is patently obvious that this is a service so valuable that it should be accepted on a nationwide basis.

It is also felt in many quarters that patients should have access to their own medical records. Unfortunately, this particular benefit was dropped from the Private Member's Bill on Access to Personal Files when it was introduced by Liberal MP Archie Kirkwood on February 27th, 1986 in order to get the Bill a second reading. He had to bow to Government pressure – a pressure, one imagines, originating from the Conservative wing of the medical profession. Yet where this has been tried on a limited scale, with maternity patients in an Inner London area and with patients in one GP practice, it has been found to work well and save the doctor time, and has been treated most responsibly by the patients.

There was a good deal of unease over the role of the receptionist. Although there were many instances of receptionists providing the most courteous and helpful of services there were all too many where this was not the case. There is no direct access to a doctor; a patient always has to go through a receptionist, so it is a job which should be treated with far greater care. There should be greater emphasis on the training of receptionists – proper formal training, that is – on how the practice is run, how to deal with emergencies, and general communication. There is a possibility that at least some of the complaints which reach Family Practitioner Committees about doctors who won't make home visits arise because the receptionist has not passed on a sufficiently clear message or because she has diagnosed the situation herself. Receptionists also need training on the necessity to check on repeat prescriptions, as is evidenced by those people now hooked on tranquillisers who have been able to keep on picking up their drugs for years without ever seeing the doctor.

The idea of a Patients' Liaison Committee seems generally welcomed by consumers. Since most practices where there is more than one doctor hold regular practice meet-

ings it should surely be possible, given the will, for this to
have some lay input.

When it comes to complaining about a doctor, there is
general agreement that the situation is pretty dire. Complaints are hard to make, the system is wrapped in bureaucracy and red tape and there is very little information given
as to how to go about it. There is virtually no publicity
given to the procedure and many Family Practitioner Committees make no effort even to publicise their own existence – according to the NCC some in the London area
are not even listed in the appropriate area telephone
directories.

At present if you want to make a complaint it has to
be lodged within eight weeks of the incident you are
complaining about. This is particularly hard for someone
who has recently been bereaved or whose relative is
seriously ill or who is confined to hospital until the time is
nearly up, especially when there is added on to all that the
difficulty of finding out how to do it. The Green Paper did
suggest a longer time limit, but this has not been received
with much enthusiasm by doctors. Consumer groups feel
that six months would be a sensible time limit within which
to make a complaint.

Complainants should be allowed as of right to be accompanied by a 'friend' of their choice. (In cases of complaints made about hospitals most health authorities will
accept a Community Health Council Secretary as a 'friend',
in a non-advocacy capacity.) There should also be the
possibility of informal methods of dealing with complaints.
In many cases an apology, an explanation of what went
wrong and an assurance that it won't happen again is
sufficient. The apology should not imply liability. However
doctors, under the shelter of their defence unions, appear
to be worried about this.

Certainly the time has come for a real shake-up of
the complaints procedure and not only at the FPC level
mentioned above where, even if a doctor is found to be
guilty of breach of his terms of service, he can often get

off with a light fine and the result of the case and its findings are published only in an official report without the doctor or his practice being mentioned.

The position if a patient tries to take a complaint even further – to the General Medical Council – is even worse. At the time of writing MP Nigel Spearing was trying to get a Bill through Parliament to alter the powers of the GMC. At present all complaints are first of all screened by a person called the Preliminary Screener. Many go no further than this. Those that get through are then considered by a Preliminary Screening Committee chaired by the chairman of the GMC. Most of these are cases which have already been dealt with elsewhere, either by the FPC or – in cases such as drunken driving or other criminal offences – in courts of law. The tiny handful that succeed in reaching the Professional Conduct Committee contain very few from actual patients.

It compares particularly badly with the Nursing Council, which takes up every complaint made to it; in fact this is mandatory under the UKCCN Act. It will do so even when the complainant does not know the name of the nurse involved. Transcripts of any hearing are automatically available (which is now not the case with the GMC) and a copy is sent to the relevant health authority to help ensure that such a situation does not arise again. At no stage is there a preliminary screening by a single individual and the committee which considers complaints has a proper input from lay people as well.

Nobody seems happy about the dental service as it is at present. There is insufficient choice, too many dentists opt only for private work, many people do not understand how much their treatment is going to cost them, and there is evidence of unnecessary dental work being undertaken to increase fees while the charges themselves are now so high they are putting people off seeking treatment at all. Resources and effort should be put into the Community Dental Service to ensure that everyone – children, the old, mentally handicapped people, and all those returned to care in the community – has access to proper dental care. The Community Dental Service needs to be built up rather than cut back as appears to be the case at present.

The privatisation of the ophthalmic services, with a very few exceptions, is now a *fait accompli*, and it will take time to see how it is working out in practice. There are many misgivings about how it will affect those with complex eye defects and those on very low incomes. While a voucher system will operate in special circumstances, this will not even cover the cost of the lowest-priced glasses and it is feared that both adults and children will go without the glasses they need to ensure adequate vision, a point made succinctly by the NCC in its response to the Green Paper.

With regard to the pharmaceutical services, it was felt that patients could be given far more information about the medicines they are prescribed in order to encourage what the Green Paper described as 'a greater responsibility for their own health'. Pharmacists should be encouraged to explain what the drug is and what it does and, where necessary, if there might be side-effects. (As for side-effects, this is a whole area in itself, too big to be tackled here. A growing number of health problems which result in patients arriving in doctors' surgeries are iatrogenic: that is, caused by the drugs they are already taking to cope with some other condition.)

General support was given to the idea of an increasingly important role for community nurses, beginning with nurse practitioners. It was generally recognised that they play a key role in health care, especially in reaching those groups most in need of it.

All in all the consumer response to the Green Paper and the Cumberlege Report was extremely positive. Most of the ideas put forward were not only practical but would not be expensive to implement. What is needed is the will to put them into practice. This requires the support of whatever Government is in power and a less conservative response from the medical and associated professions.

But the consumer bodies were not happy about the

present attitude to the overall funding of the NHS, either at the hospital end or in primary care, nor about the importance given to either preventive medicine or health education. Both are vital; and not only are they closely interlinked, but it is very difficult to get the message of their importance across to those very groups most in need of them.

8

Positive Health

It is apparent that far more could be done to prevent sickness and promote positive health. In an ideal world we would make certain that adequate resources were made available to ensure that no one need be badly housed, badly fed or without a sufficient income. We tend to forget that the great breakthroughs in public health came about not only because of the discovery of vaccines and better drugs, but also as a result of simple public health measures.

It was because Dr John Snow noticed that the 1854 London cholera epidemic had started with those who drew water from a pump in Broad Street, that the pump was closed down, and that source of infection stopped immediately. It was apparent that a clean drinking water supply in tandem with a proper sewerage system would halt a good deal of disease without recourse to medical treatment at all.

Gradually, over the years and usually with positive antagonism from the industries concerned, it became apparent that a whole variety of diseases from pneumoconiosis to mesothelioma (a cancer of the lining of the lung found among asbestos workers) were caused by the work people did. Increased health and safety regulations and the prohibition altogether of the use of some substances has helped to prevent at least some industrial disease.

It is obvious therefore that decent housing would drastically reduce the number of those suffering from, for instance, respiratory diseases, a good diet those suffering from obesity and heart trouble, and jobs for all those suffering from the mental and physical stresses caused by prolonged unemployment. However the National Health Service cannot provide housing, social services, millions of much-needed jobs and a nationally agreed sensible diet.

But there is much that could be done, even so. The principle of screening for the early detection of disease has been long established, and there are many of us who can remember the arrival of the chest X-ray van which invited you to have an X-ray at a convenient local point. Before modern drugs made it a rare condition, there is no doubt that the chest screening picked up tuberculosis at an early stage along with other chest troubles. No one would now disagree that cervical screening is also necessary to pick up cervical cancer while it is still easily curable but, as we have already seen, cut-backs have affected every stage of the cervical cancer screening process. About 2,000 women die needlessly of cervical cancer every year. At the time of writing the Government had just announced that, after all, some extra funding will be found for cervical screening along with money to start a national programme of screening for breast cancer. This will be limited at first to the 15–64 age group. Every year 15,000 women die of breast cancer. It is the most common cancer in women.

We have waited a long time for action on screening for breast cancer and greater resources for cervical screening. Those cynics among us would say that if as many men died of cancer of the testes, because it was not diagnosed in time, as women do of breast and cervical cancer, there would be a screening clinic on most street corners with a special one at the House of Commons.

But unless screening is treated practically it will still be the articulate predominantly middle-class women who will benefit most. They will push their doctors to give them what they want and put pressure on their local health authorities; when it comes to the basic business of getting to a hospital or clinic to be screened then they will probably be able to do so by car. For the women living in the middle of wasteland housing estates or in country villages with little public transport and with no access to a car, a visit to a screening centre might mean a protracted and difficult journey, often accompanied by several small children. In the end it all becomes too much trouble. Surely, therefore, we should be thinking again in terms of the mobile screening service which could park itself on housing estates, in villages (much as the county library service does) and in the deprived inner city areas? There have been one or two pilot schemes offering mobile cervical screening and these have shown an excellent take-up.

In fact there is now considerable support for the idea of Well Women's Clinics or Centres, and a few are actually operating. One of the first Well Women's Clinics, as it was then called – now the term 'Centre' is preferred – was in Islington, followed by another in Manchester, both with similar goals but operating slightly differently.[1] From fourteen to ninety-four any female who presents herself to her GP with what might be termed a female ailment is likely to be told 'it's your age' or 'you'll have to put up with it'. Women do have special problems connected with menstruation, pregnancy and all that might happen during it, and with the after-effects of childbirth and the menopause. The problems can be slight enough merely to be a bit of a nuisance or serious enough to make it difficult for the woman to cope. As in most cases the woman is still the linchpin of the family it is obviously worth making sure that she is as healthy as possible so that she is up to meeting the demands made upon her.

Again all the emphasis should be on positive health. The Centre itself may or may not offer treatment, depending upon how it is staffed, but would offer medical screening in the context of health education and counselling and a whole range of services currently under a number of different roofs, e.g. family planning, chiropody, etc. It might also offer full check-ups and screening for conditions such as sickle cell anaemia and thalassaemia in areas where there are ethnic groups prone to these afflictions.

It would offer advice on diet and weight, and make use of lay personnel for help with counselling. When a conference on the subject was held in Manchester some years ago women felt that this was an area much neglected – women who had experienced miscarriage, Caesarian section, mastectomy, hysterectomy or had gone through the menopause might well be able to help other women to face similar situations.

Serious conditions picked up during the overall health check-up or screenings would, of course, be referred either to the woman's GP or, if necessary, to a consultant.

To attract as many women as possible to make use of such a service ideally centres would be housed in shops in busy precincts and would offer a walk-in initial interview. For women living on estates or in rural areas, then again there should be the possibility of a mobile service. Since many of the services which such centres should offer are already provided but not under the same roof, consumer groups felt that the extra cost to the NHS would be minimal but the saving on bad health appreciable. Although some health authorities have gone a little way towards meeting the principle of the Well Woman Centres it is obviously apparent that there is still a long way to go.

From this idea developed others such as Well Pensioner Clinics, designed to help the elderly keep fit and active as long as possible. With these, as with the Well Woman Centres, real efforts would be made to attract members of minority ethnic groups to use such facilities; they often have special problems and are even more reluctant than most to use the NHS – whatever those who like to criticise them may have to say.

The ideal would be attractive Well People's Centres. A

dream? Maybe but, as we will see in the next chapter, it
has actually been tried.

It is almost impossible to discuss preventive medicine
without also bringing in health education. One of the first
priorities of any new incoming Government must be to
bring back the Health Education Council. The new government body whose work will almost entirely be concerned
with fighting AIDS does not really replace the Council.

No one who saw the *This Week* programme on health
in Sheffield which was mentioned earlier could be less than
appalled at some of the interviews. Unhealthy-looking
women, coughing away, were seen chain smoking while
being asked how they felt about the high level of early
deaths from lung disease.

They seemed quite defeated. 'Smoking's my only
pleasure,' said one. 'I don't go out no more with my
husband on the dole. What else would I have to live for?'
Yes, they knew it was bad for them. How about screening
for cervical cancer? Yes, they knew about that, but if you
were working then you hadn't time to go up to the hospital
and if you weren't – well you couldn't afford the bus fare,
could you? Even, apparently, if you could afford to smoke.

This confirmed the findings of a survey into what people
thought of the Health Education Council's booklet 'Beating Heart Disease'. In *The Politics of Health Education* by
Wendy Farrant and Jill Russell similar quotations are
given, such as:

Everyone knows smoking damages your health . . . it
doesn't bother me that I smoke. It did when I was on
the pill and had high blood pressure.

I think smoking does pull you down but when you've
got three kids running around the house it's ridiculous
trying to stop.

. . . if ordinary people read this kind of thing and see
they have to keep their weight down and stop smoking,
there's nothing to live for.

My husband was on a 2,000 calorie diet before he died of heart disease . . . he had to eat salad – it takes all the pleasure out of living. Better to have a shorter life.

Butter's the main part of your life. What's a chip butty without the butter?

Why make life miserable? You might as well have a shorter and happier life.

When asked further about the causes of ill health the respondents agreed that smoking and diet had some part to play but were also well aware that this was not the whole story – there was far more awareness than one might imagine of the possibility of pesticide residues in food and the effect of additives, which was interesting. Also they saw an obvious correlation between poor health and poor housing, unemployment with all the stress and worry that brings 'this time is more worrying than others, it creates lots of illnesses . . .', the Western way of life and the need, if you are in work, to keep up the payments on the mortgage, on the HP, etc., and the added stress that also brings.

Yet what seems to be almost universal in all these surveys is that those surveyed seem either apathetic about the situation or do not see that they themselves can do anything to better it. A Social Science Research Council survey as long ago as 1975 discovered, when there was substantially less unemployment, that people put their health as a low priority when considering what made up their 'Quality of Life': family, home life, marriage, money and the cost of living were generally considered more important than how healthy they were.[2]

On top of all this too, of course, comes the enormous power of advertising, the advertising of all those things which contribute to poor health. The average family watches hours of television a day – in houses where bread-winners are on the dole then it is rarely switched off. From dawn to midnight they are bombarded with messages:

'the snack you can eat between meals', the sugar-coated chocolate tit-bit, 'the lady loves Milk Tray', the pure countryside which comes into your home when you eat butter, the sophistication of drinking Martini, the instant soups and instant sweets, the happiness of a cigar called . . . we can all finish the sentence.

Faced with this, plus the further advertising in the media, on hoardings, in the underground, on buses, advertising material which sets out to try and inform of the hazards of too much drink, poor diet, smoking and lack of exercise becomes just another message to be taken in alongside the far more massive onslaught of those whose profits depend on heavy sales of those very items people are being warned against.

It is against such a background that the busy GP in his surgery and the community nurse on her rounds are supposed to be devoting time to preventive medicine.

The World Health Organisation in its 'Charter for Action – Health for All By the Year 2000' sets out the roles that should be played by governments, local authorities, health professionals, trades unions and ordinary people.

On primary care it says:

Primary health care is essential health care based on practical, scientifically sound and socially acceptable methods and technology made universally accessible to individuals and families in the community through their full participation and at a cost that the community and country can afford to maintain at every stage of their development in the spirit of self-reliance and self-determination. It forms an integral part both of the country's health system of which it is the central function and main focus, and of the overall social and economic development of the community. It is the first level of contact of individuals, the family and the community with the national health system bringing health care as close as possible to where people live and work, and constitutes the first element of a continuing health care process.

Leading out of that comes the individual's response.

It is people who ultimately decide on the value of health in their lives, although their real options may be severely restricted by the economic, social, cultural and physical environment. People have the right to be informed. Collectively health professionals share the responsibility for broadening the framework traditionally used to define and analyse health problems by looking more into the psychological, social, economic and environmental determinants of health . . . and stressing the importance of acting on those determinants if health is to be improved. Health professionals should also help in making the facts in this regard better known to the public . . .

The response of the present Government was to disband the Health Education Council. A prophetic statement was made to the authors of *The Politics of Health Education* by a spokesman for the HEC before they knew it was to be disbanded. He/she said:

> . . . it is to the discredit of our message that we get constrained either by government or commercial bodies. The great strength of the HEC should be that it is seen as only interested in providing a health message. It should not be constrained by government views and it should not be constrained by commercial interests . . . When it gets embroiled and either ends up being bought off by companies or being seen as the government's mouthpiece – as some sort of government health propaganda – the HEC will lose credibility.

Amid so much gloom it looks as if it is almost impossible, even if there were more resources, to provide a preventive health service coupled with health education especially to those who most need it. But in fact it has been done – once. It was called the Peckham Experiment and it worked.

9

The Peckham Experiment

In December 1985 there was a celebration to mark the fiftieth Anniversary of an experiment in positive health which had been light years ahead of its time. There had been nothing like it before and, sadly, there has been nothing like it since.

The 'Peckham Experiment'[1] grew out of the work of two doctors: Dr George Scott Williamson, formerly a pathologist at the Royal Free Hospital in London, and his wife, Dr Innes Pearse, who had been a medical registrar at the same hospital. During their work, which had brought them into contact with many families from deprived social groups, they had become convinced that health in the family started before birth. Prospective parents should be in the peak of health before conception, they should want the child, and they should be fit and eager to rear it. How could families be encouraged to be healthy, particularly in what we would now call the deprived inner city areas and in the low-income groups?

The beginnings of Peckham were a small combined health centre and club set up in a house in south London. A group of people who were interested, virtually all of them under thirty, got together to organise it on the premise that families needed to be offered a health service in the real meaning of the term. The club would aim to provide a service which would enable local families to ward off sickness before they were smitten with it – wherever possible – and this would be done by means of periodic health overhauls for all its members with ancillary services for infants, children and parents alike.

The 'club' operated from 1926 to 1929 by which time it had vastly outgrown its accommodation, so the next step was to raise the money to build the best possible kind of health centre. The imagination behind what was to become the Pioneer Health Centre has never been matched. It was far ahead of any kind of so-called health centre we have today.

The Centre was designed by Sir Owen Williams. It had glass walls and a splendid swimming pool in the middle of it, at that time the second largest in London. There was a theatre, a gymnasium, a children's nursery, a cafeteria, a library and a suite of medical rooms.

It was a club for the whole family and the whole family had to join. It cost them 1s. (5p.) a week. The conditions and privileges of membership were two:

1 every individual of a member family had to agree to a periodic health overhaul;

2 use of the club and all its equipment was free to all children of school age or under, from each family, and by the adults on payment of a nominal additional charge for each activity.

As the doctors themselves put it, 'a whole lifetime may be spent in the process of dying. "Survival" isn't living; nor is health the mere absence of disease.' So instead of studying health in the context of disease, they resolved to study disease in the context of health.

Once the health overhauls got under way the doctors discovered that even those people who thought they were healthy were very often not so. Out of 3,911 people in families who were examined in the annual health overhaul over 90 per cent were found to have something wrong with them. This was often something quite minor like poor teeth or a bad skin but sometimes it was far more serious

and the Centre turned up the kind of diseases so prevalent
in deprived areas today – heart disease, chest and respiratory diseases and cancer. People found with such serious
conditions were referred to ordinary doctors or hospitals
for treatment.

Well ahead of its time – again – the Centre offered
family-planning advice, but it also encouraged women to
have babies and some women who had either decided
against having children at all or who had decided to have
no more, changed their minds after attending the Centre.
When they became pregnant women had a special consultation session where the course of the pregnancy could
be discussed along with diet, problems arising from the
conditions in which the woman lived and the conduct of
labour and delivery. The husband was positively encouraged to attend this and subsequent sessions. The woman
was then seen every two weeks throughout her pregnancy
to check that all was going well.

Diet was considered to be of the utmost importance and
the family overhauls showed how inadequate this often
was. The family were advised on the need for a healthy
diet but it was also obvious that in such a low-wage area
eating well was all too often a doctrine of perfection.
Because of the need to ensure a healthy baby, therefore,
the pregnant woman should be given every assistance to
eat well; to this end the Centre bought a farm near Bromley
in Kent. It had a herd of milk cows and produced organi-
cally grown vegetables and both milk and vegetables were
then sold to pregnant women at the Centre at a basic price.
It was also discovered that most women suffered from an
iron deficiency during pregnancy and so supplementary
iron was given when necessary.

Just as forward-looking was the Centre's attitude to the
delivery of the baby. The presence of fathers was not on
the agenda in those days, even at Peckham, but they were
encouraged to take an interest. Ideally, said the Peckham
doctors, women should be delivered at home if there were
no complications, but they recognised that this was often

not possible where the women were living in small, cramped flats with no room for other children to be elsewhere, often with only a cold tap and a lavatory shared with other families.

So an arrangement was made whereby a Centre midwife looked after the woman until she went into labour, then, following an agreement with a local hospital, the woman went in to deliver her baby, remaining there only for forty-eight hours. She would then return home to find the Centre midwife waiting for her and her postnatal care undertaken by the Centre. This is almost like today's 'domino' system, where a midwife cares for a woman during pregnancy, accompanies her to hospital to deliver the baby, then continues her care when mother and baby return home after delivery.

The Centre monitored babies frequently, continuing regular check-ups of the child right through until school age, when it then became part of the general family overhaul system.

The first years of the Centre were from 1935 to 1939 and then the premises were taken over for war service. It then reopened in 1946 and ran until 1950. By the latter period it offered all the services already mentioned plus a nursery school, youth club, marriage advisory council, Citizens' Advice Bureau and child guidance clinic.

Writing in the early days of the Centre in the book *The Peckham Experiment*, Dr Innes Pearse described a day in the life of the Centre from the afternoon when male shift-workers would come in to play billiards or swim in the pool to a dance at the end of the evening. During that time young women would come for antenatal check-ups or to have their new babies examined, younger children would go to the nursery while other women attended Keep Fit classes, and yet other women were making children's clothes using the Centre's sewing machines.

In the gymnasium area couples would be playing badminton and, after school, children using the equipment there or swimming in the pool. The cafeteria would be in

full swing offering wholesome, but attractive, food at a
basic price. A rota of women assisted in the cafeteria with
'nursery teas' for toddlers.

There was generally a lull around 6 p.m. as women and
children went home and the men arrived back from work,
then the activities started up again – in the swimming pool,
the gym, the dance hall, the theatre where a group might
be rehearsing the next play, the billiard room, the games
room where darts and table tennis were in full swing, and
the cafeteria again full. It was licensed and sold beer as
well as soft drinks and food.

On an average evening by 9 p.m. there were between
500 and 1000 people in the Centre, aged from fourteen to
ninety . . . And while all the general activity was taking
place, some families were going through the procedure of
the family overhaul. After the results of this had been
assessed the families would be called in and have the
findings explained to them. They were encouraged to ask
questions and problems were sorted out, advice was given
and, when something was found to be badly wrong, the
necessary referrals made.

But it was not just the people of Peckham who learned
from the Centre. The doctors learned from the people. As
Dr Pearse says:

Our failures during the first eighteen months' work
taught us something very significant. Individuals from
infants to old people, resent or fail to show any interest in
anything initially presented to them through discipline,
regulation or instruction, which is another aspect of
authority. (Even the very 'Centre Idea' has a certain
taint of authority and this contributed to our slow recruitment.) We now proceed by merely providing an environment rich in instruments for action – that is, giving a
chance to do things. Slowly but surely these chances are
seized upon and used as opportunities for development
of inherent capacity. The instruments of action have one
common characteristic – they must speak for themselves.

The voice of the salesman or the teacher frightens the potential users.

So why was it that this excellent experiment in positive health did not lead the way when the NHS was finally set up? Why indeed. As Colin Ward, one of the many commentators who have written about Peckham put it in his article in *New Society* on December 20th, 1985:

> . . . since 'health centres' had become part of the official doctrine after the passing of the NHS Act in 1946, the directors [of the Centre] approached the Ministry of Health to find a formula to incorporate it into the official provision. They failed for five reasons.
>
> 1 It was concerned exclusively with the study and cultivation of health; not with the treatment of disease.
> 2 It was based exclusively on the integrated family; not on the individual.
> 3 It was based exclusively on a locality; it had no 'open door' policy.
> 4 Its basis was contributory (by 1950 it was 2s.6d. a week – 12 and a half pence), not free.
> 5 It was based on autonomous administration and so did not conform to the organisational structure of the NHS.

That what it was attempting to set out to do was not recognised is tragic. It promoted positive health in all its aspects and it encouraged people to take responsibility for their own health, with ongoing assistance from the Centre. It saw health as being more than the absence of disease. It could, though, pick up disease at an early stage and it could also help its members with their psychological and social problems. It was a social centre where the whole family could go and meet other families. It encouraged healthy diet and healthy living without adopting a proselytising approach. At the very least, attempts should have been made to monitor the families involved to see how they fared later and how the Centre affected the later health of the children born under its care.

The refusal to consider seriously what Peckham was trying to do was a dreadful missed opportunity. Those administering the NHS learned nothing from the message of the Peckham doctors when considering community health – that a paternalistic, authoritarian approach to promoting personal good health does not work.

As Colin Ward says:

We failed to grasp this when we conceived postwar welfare in terms of bureaucracies and hierarchies. Nobody in authority thought there was any importance in the Peckham concern with the sources and origins of spontaneous action. Can we at last learn the patience and humility to accept that the only kind of social organisation worth striving for is community self-organisation?

10

Second Opinions

How do other people see the way forward for the National Health Service? To canvass other viewpoints I have gone to four specialists, two lay and two professional.

Jean Robinson was the second chairwoman of the Patients' Association, is currently a lay member of the General Medical Council and was a member of Oxfordshire Community Health Council. She is also involved with a number of other organisations such as the Association for Improvement in the Maternity Services (AIMS) and a women's refuge.

This is her view of the way forward:

I think that the main problem in the NHS is that it has always been profession-dominated, mainly doctor-dominated. They always made the policy, the decisions, called the shots, and there has never been any counter-balancing power for the consumer (although whether you use the word 'consumer' is a moot point – you might have a choice of which brand of baked beans you buy but you don't have a choice as to whether or not you get appendicitis . . .). But we are consumers certainly in that we pay for the whole thing. Yet there has never been any allowance for this – such as a fair complaints procedure, with regard to either the general practitioner or the hospital services, nor has there even been a system whereby complaints can be constructively used in the way a company doing market research would use them to say 'what's wrong with our product?'

This is because we have never paid at the point of sale or had democratically elected Health Authority Committees or Family Practitioner Committees or so on. The whole way it was set up was geared against us, for instance the system whereby we have to register with a general practitioner which leads to difficulties when moving, if you want to change – and we know of areas where doctors have a gentleman's agreement not to take patients from another area – and if you make a complaint. Now if you take the French system, and I'm aware that it, too, has its problems, you don't register with one GP, you go for a consultation with the doctor of your choice and you pay and then either claim most of it back from the State or if you are exempt, then all of it. But he knows that his income depends upon you choosing to walk through his door in the first place and you choosing to walk back through it again.

Now here the majority of the doctor's income does not even come from the annual fee for each patient registered with him. More than half comes from all the allowances given by the FPCs, so he does not see his patients as a direct source of income. Consequently, we have neither the benefit of a market economy (I don't think there is any such thing as a real *free* market economy in medicine) nor the benefits of choice or democratic power.

By and large it does mean that most doctors are reasonably caring – certainly in comparison to, say, the USA which is mostly private practice. We have had a reasonably responsible and reasonably conscientious profession; but we have also had a system where those who are not responsible, conscientious or adequate have been able to get away with it for a large proportion of their careers.

The NHS has also been profession-dominated in the kind of choices that have been made, the kind of things that have been made top priority like acute medicine, 'high tech' birth, heart transplants, and so on, and even the kind of prevention they have thought necessary, such as accepting the need for cervical screening but not giving women adequate information on how to avoid getting the disease in the first place. I feel most angry about what has happened in obstetrics where the midwife, the low-cost supportive midwife, has had the rug pulled out from under her and has become virtually powerless; where even in her professional capacity she is told to carry out certain procedures whether or not she agrees with them.

What I feel very depressed about is that even where you have very well informed and highly responsible consumer pressure there are such difficulties. I'm thinking particularly of obstetrics and groups like the Association for Improvement in the Maternity Services who have really done their homework, produce an excellent journal, offer supportive help to women, pick up the pieces emotionally and deal with problems of sexual dysfunction following a bad childbirth experience, groups who do excellent therapeutic as well as pressure-group work and can offer a great deal of expertise and information, but make virtually no headway against the absolutely entrenched power of the medical profession. Of course there are the individual doctors who have been willing to listen, like Wendy Savage and Peter Huntingford, but who then come under great pressure from their colleagues; but overall the attitude has been depressing.

I have been working in the consumer aspect of health care for twenty years now, since I was put on a Regional Hospital Board as the token housewife. During that time, although there have been changes and improvements, I feel on the whole pessimistic about it all, as if the power of the medical profession is such that we are never going to have any real impact even though we are paying most of their salaries.

I think medical education is appalling but how do you effect any kind of radical change in medical education? That goes right through the whole of medical training and, of course, the Royal Colleges dominate medical education. I will say that the Royal College of General Practitioners has been far more forward-looking than the rest, in having a patients' group and involving them in policy-making and so on, and this is obviously a good trend.

But I feel we in this country are one of the most appallingly ill-educated populations when it comes to health care in the industrialised world. Compare us with America. Americans are not, of course, necessarily better cared for than we are, as doctors there are paid on a piecework basis, which is a deeply dangerous way to pay doctors. But the level of ignorance among people here about what high blood pressure means, what hypertension is, what your kidneys do, is astonishing. I met this again and again with the Patients' Association. It makes me wonder what on earth our schools are doing. I suspect the level of education on the physiology of our bodies has actually deteriorated. I left school just after the war and at least there was plenty of education on nutrition in those days and on keeping healthy.

So we have this appallingly ill-educated public who are then criticised for making bad choices and wrong decisions at the same time as they are being bombarded by advertisements like 'a Mars a day helps you work, rest and play' – what does it actually provide you with, for goodness sake! It is this kind of thing that people remember, not the more complex information on health care.

In a way – and you might think this is an odd thing for me to say – the fact that we did not have to pay at the point of contact with the health service and therefore felt we were getting it free (even though we were paying and are paying) meant that we have always felt beholden to those who provided the service. It's amazing how passive and unwilling to criticise we are. Because of all this people have been less well educated than they otherwise would have been. In America they are mal-educated but they are not uneducated.

They go overboard on vitamins and investigations and so on, but if a doctor is prescribing a very expensive medicine which you or your insurance company has to pay for then you really want to know whether you are getting value for money.

The other thing that has militated against better health care is this lack of choice. Usually you register with a GP and unless he dies, or something really awful happens or you move, then you are stuck with him, so a large number of people never know what good general practice should be like. They are bound to think that what their particular doctor does is what all GPs do. So I've come across people who have gone to their doctors year in and year out and have never had their blood pressure checked or been properly examined, and have had repeat prescriptions for valium or sleeping tablets or been given antibiotics where these were not in the least indicated and this, to them, was medical care because they have had no opportunity to compare one practice with another. This is not just evidence coming in from pressure groups but from the medical press, where doctors themselves have taken locum jobs and found surgeries without even the most basic equipment and where there are no adequate notes on patients.

In fact we have had a system which has allowed really lousy standards of care to exist not just in respect of actual danger to patients, but in behaviour towards patients which is completely unacceptable and by people who are paid by the taxpayer. Some of them turn up in the Ombudsman's Reports, especially consultants, yet apparently everyone felt powerless to do anything about them, and where, apparently, the administrators of the service felt it their job to defend the staff against complaints not to act as watchdog on behalf of those people who pay their salaries too. There must be a better system.

Now regarding health education, to blame people for what we see as unhealthy habits without taking into account the huge amount of money spent on advertising, I think is ridiculous. We now know, for instance, that most people who start smoking do so in their early teens and the habit is established before twenty. So we really have to look at the pressures which start kids smoking in the first place. If we can stop them before they are twenty, we can stop them – period.

And I think we should be honest about other causes of ill health as well as smoking and drinking. For example if you smoke but are in a white-collar job and living in an unpolluted area like East Anglia and having a good diet then your chances of getting lung cancer are increased but are nothing like so great as if you live in a bad area and work, say, in a foundry. Then you are signing your own death warrant. Mind you, by working in heavy industry anyway you are increasing your chances of death. So you can actually see from figures published by the Registrar General that there are vast differences in the death rate of those who smoke, depending on how they live. From looking at figures of people who smoke in other countries we can see it is not just that on its own – although obviously I'm not suggesting that anyone should smoke. The conditions in which people live should be taken into account by health educators; they shouldn't just simply blame people.

Nor do they take into account other factors. When you see women with kids around them who are smoking, you feel sad for the women because so much of their budget goes on it – often straight out of social security – and so goes right back to the government.

If you talk to those women, what are you going to replace it with? Somehow smoking is the only thing they have which is theirs, for *them*, that comes out of the household budget because they are addicted to it. You feel that if they weren't smoking they wouldn't be having little treats like going to the cinema or the hairdresser, they wouldn't have anything. It seems to me that it is all very much bound up with the quality of women's lives. It's even worse when you talk to schoolgirls, as I do – they say what do you mean we're going to get lung cancer at forty? We don't expect to live until we're forty, we'll die in a nuclear war before then. Culturally they live in a different age from us, where the likelihood of nuclear disaster is so much more important than that of lung cancer.

Then comes the question of a healthy diet. I work at a women's refuge part-time and I'm very concerned about the nutritional state of women. Usually a woman in the family will look after herself last. When she is depressed she doesn't usually have the energy to bother feeding herself properly. A lot of food education talks about the nutritional values of certain foods but what it does not take into account is the energy content – not the energy content of the food but what might be termed its labour content, the amount of labour needed from the woman to prepare it. So jacket potatoes are fine – but they have to be lugged back from the shops, say, by a woman with four kids. Imagine the labour involved in bringing back from shops, which might not be near, the amount of vegetables, wholemeal flour and so on needed for such a family. So it is the labour of shopping for it, physically transporting it back and cooking it and the fuel costs of cooking it properly. On the whole I think those who give out nutritional advice think in terms of middle-class women when they speak of vitamin and fibre content – sometimes they think of price – but they just don't consider the 'women's labour' content.

They should consider the real circumstances, the energy levels of women who are not well and/or depressed and I get very angry with all this moralising over convenience foods. In the medical profession there is a huge literature now about 'burn-out', burn-out of doctors, nurses, the health care providers. What about burn-out in carers who have a twenty-four hour day, seven-day-a-week caring job and no time off, no pay, no holidays, such as unsupported mothers? I think an enormous amount of effort should be put into finding and then pushing convenience foods which are cheap, nutritious, light to carry home and easy to prepare. Fish fingers are a good example – now, alas, rather expensive – but fish is a very good food and kids will often eat fish fingers when they wouldn't eat fish.

So basically I think a lot of health education ought to be going into saying if you haven't got the time and energy to cook and are feeding a large family then eat what is the easiest as well as the most nutritious, for Mum's energy is valuable. There's far too much unreal moralising now about how every woman should be making lentil soup, muesli and her own bread!

Tom Richardson is the Secretary of Oxfordshire Community Health Council, a post he has held for eleven years. He is one of the most experienced Community Health Council Secretaries and is also a county councillor.

Mr Richardson says:

The Health Service in my view badly needs some democratic input. What we have at the moment is chairmen of health authorities appointed by the Secretary of State and paid an honorarium by him of around £10,000, with general managers beside them who leave little room for any public input particularly at regional level, where people might have to drive seventy or eighty miles to a monthly meeting knowing almost nothing about the problems the same distance the other way, from which other people have come to the same meeting.

So it seems to me that members of all these committees should be democratically elected in some way and that we should be talking about how to do it, not whether to do it or not. That's axiomatic. Otherwise even if we have a Labour Government that starts pumping money into the NHS all that will happen is that there will be more people in more hospitals. We don't need them. It does nothing for primary care, it does nothing for health education and nothing to help keep people in their own homes in the community. So what I would like to see being discussed is how to make these committees – and indeed Community Health Councils for that matter – democratically elected and accountable.

There's no substitute in my view for such a system. In Oxford City we have an election every year for one third of the city council and every fourth year the whole county council comes up for re-election; those of us who are involved have to spend six weeks on the doorstep defending our record on the council and that is good for the soul. If people had to face up on the doorstep to working people who are kept hanging about in out-patients' for hours at a time, people who are crippled waiting for operations for weeks if not months, in danger of losing money or even their jobs because some unsym-pathetic GP's receptionist won't give them an evening appointment, then I think we would see some real changes. With regard to GPs, some of them are actually doing away with evening surgeries altogether and not even having Saturday morning ones.

We have the situation now where there are not enough GPs in inner city areas, yet I was lecturing to medical students only this week – there were about fourteen of them – and they told me they had heard it was unlikely they would get a job. They should be salaried. These were excellent well-qualified young people and they should be offered jobs immediately. We should be think-ing of setting up a State-run general practice service. I think the two things could run very comfortably side by side, it would give people choice. For instance there are many towns in Oxfordshire where there are supposedly two practices or even three but in fact the GPs have a restrictive practice agreement that they won't take one another's patients.

So there is no choice. But a really good health centre run by properly salaried caring people would certainly smarten up the individual contractor service of the other GPs and I think this could be relatively easily done and not terribly expensive. Especially if you involved the local authority in setting up health centres again.

Before the 1974 reorganisation there were many local authorities who had a good record in providing health centres in areas where they thought they ought to be. They might not have been the kind of health centres that we all might like – such as Peckham in the 1930s – but they offered a good basic centre and there have been none of them built since the local authorities were pushed out in 1974. For instance in Oxford there's a new estate going to go up soon of over 1,000 houses. Now prior to 1974 the planning of that would have included among other things a health centre, but now that doesn't happen and they are wondering how to get GP input into it. People stood a vastly greater chance of getting a health centre prior to 1974, by approaching the appropriate committee of their local authority on which sat local people with local knowledge. They could have either told enquirers why they couldn't have it because of where it came in their list of priorities, or said yes, certainly now let's plan for that.

Now we have the additional problems of people coming out into care in the community, coming out of the responsibility of the health services and into local authority care. Local authorities on the whole have an excellent record in health care. The major changes in the health of the population have come about because of the local authorities – better housing, better sewers, a better water supply. They pioneered primary care teams – for example Oxford City was the first authority to attach district nurses and health visitors to GPs, and that was in the early 1960s. It didn't come about through the NHS. That was then copied throughout the country.

So to me local democracy has put an enormous amount into good health and I'm really not at all sure how much the NHS has provided with its vast hospitals and high tech. At the moment there is considerable pressure for the notion that the NHS should not provide any continuing care and this is growing. This is shown by the number of health professionals now becoming involved only in private nursing homes, the number of closures of hospital beds for the elderly who might have to spend their last weeks or even months in them, and yet without any real planning of transfer of money from the health service to the local authorities. This leaves the voluntary bodies to cope with what will be an increasing number of elderly people – and this has all happened without public debate. At least the closure of the large mental hospitals took place against some background of public debate.

I am deeply worried about care in the community for the elderly. Geriatric hospitals are closing all over the country, some general hospitals treat people over seventy-five as if they were pariahs who should not be taking up beds, GPs in London are being reported as striking people off their lists when they reach seventy-five, there are consultants referring elderly people only to private nursing homes in which they may or may not have a financial interest. So it all seems to be pointing to a new attitude within the NHS towards the elderly. If you are over seventy-five and need acute treatment you might get it on the NHS, but if you need anything other than that then you'll have to pay for it yourself.

Overall I think the NHS is a splendid example of how grotesque it is to try and run things centrally. There has been over forty years of total over-centralisation from the DHSS at the Elephant and Castle through the fourteen Regional Health Authorities in England and yet the poor still get the poorest treatment, the best areas of the country still get the best. Nothing has changed in forty years except that, hopefully, the scale of care has at least moved upwards. This appalling centralisation has meant that local treasurers haven't even been able to invest petty cash – which comes to considerable amounts a month – into a bank to get the interest on it; in fact they are prohibited from so doing. Compare that with the amount of money local authority treasurers earn for their councils by careful investment where it can earn 8, 9, 10 per cent a year in safe things.

So you could make an enormous and immediate change by saying that such and such a health authority can spend £110 million in 1987/8 and then see that they got it on April 1st or even that it was divided into four equal parts and they knew that they would get each quarter reliably on a set date and that they were allowed to invest. This would change the face of the health service considerably.

There are those that say that health services should come under the aegis of regional government if we were ever to have it. I'm in favour of that and of regional government provided the power was clawed back from Whitehall and given to the county, district, town and parish councils. Labour Governments have not been at all good in this respect. They have not pushed their civil servants to relinquish power to the regions. So if you had proper regional government and not just junior regional councils responsible only for trivia, then I think there would be a case for making them responsible for the health services in their areas as well.

Dr Patrick Pietroni is chairman of the British Holistic Medical Association, which he helped to found, and Senior Lecturer in General Practice at St Mary's Hospital in London. He has been carrying out research at the Lisson Grove Health Centre and is to be the head of the new Wates St Mary's Health Care Research Unit.

In his book *Holistic Living*, Dr Patrick Pietroni defines

what is now becoming the positive view on health. He offers seven propositions:

1 Health is everyone's business and should not be left to doctors.
2 Health care involves ethical, moral and political decisions on the part of planners and doctors.
3 Patients should assume more responsibility for their own health care.
4 Illness, disease and pain, suffering and death are all part of the experience of being a human being and we cannot escape from these experiences.
5 Promotion and preventive health care is cheaper, more effective, and less dangerous than curative and restorative therapies. The meaning of an illness may be sought on many different levels.
6 Drugs and surgery do not allow us to care for or even cure many of the current concerns of humankind.
7 Low technology, primary health care delivered within the community should receive a higher priority than it now does.

He is at present setting up a new kind of practice. It will provide the normal GP services but also offer alternative therapies where this is felt to be appropriate, and will also feature a major educational and promotional programme. He says of this:

Hitherto I feel that has always been seen as the icing on the cake. There has been shortage of funding, shortage of personnel, shortage of energy and shortage of creative thinking. In our practice the educational and promotional programme will be inherent in the work. For instance if I see someone with high blood pressure, say, or eczema I will say to them: 'Look, before I see you next time, I should like you to have attended these classes which will help you to deal with some of the symptoms you have' – so that the whole emphasis of the practice will be far more on a shared partnership. The doctor will not be there as the one who will cure the illness, he will be there much more as a coach or guide. It will be for the patient to take over some of the responsibility. The NHS has been far too doctor-dominated.

But there is another whole group whose circumstances and background are such that the concept of taking responsibility for their own health is totally inappropriate – they are at the level of basic survival. I don't think it is right to adopt a model of health care which is not appropriate to that group. So I think there always will be a place for the interventionist approach, the more conventional and caring role that the general practitioner has always maintained.

Basically I think it's a question of different health needs for different parts of the population. Some of those attending the practice will want to take a more positive part in it, will want to attend the classes, etc., while others will want to see the doctor in the more traditional role. What is required is flexibility to accommodate the differing needs of a population, not to say 'this' is the right way. It's a question of blending, of having under one roof a whole range of activities; so you will have 'high tech' medicine, you will have drugs, you will have scientific work and investigations, but you will also have an educational approach, a sense of sharing, or understanding, of learning with the patient. The role of the practitioner then becomes that of a guide to finding the model which is more appropriate for that particular patient. Because we have quite a large research programme, we will be able to evaluate models of health care.

So that's one aspect of primary health care. The second major departure from the present system of primary health care is that up until now primary health care has mainly been response-based; i.e. you have gone to your GP when something has gone wrong. So there will also have to be what one might describe as more of a community sense within the practice which is something we are trying to develop amongst the practice population of patients, clients, consumers, shareholders – whatever word you use – a sense of it being *their* centre.

There have been past attempts like patients' groups and users' clubs and so on and I think we will try to learn from those attempts; but we will be working towards a situation where the practice population becomes its own support group, because within the practice we may well have many people who will be willing to give one or two hours a month – very much like the Peckham Experiment. In a way it will be a kind of second Peckham, 1980s style, having learned a good deal from that. So there will be a community approach model where we will try and have a monthly meeting to share our problems. We really will be asking and seeking their guidance in running the practice and looking for their help in offering a network of support rather than on having to rely on a rather depleted and under-funded NHS.

Yes, there is a need for primary health care and the NHS provides the base on which to develop, but I think it needs to move away from the notions of 1947 of the doctor-centred medical service towards a more client-shared and participatory one.

However this does seem to require a very different outlook from the medical profession itself, and as Dr Pietroni is also a lecturer in general practice I asked him how this was being tackled.

It is being tackled, but it is a long haul. Changes are taking place. I don't think it is totally the doctors who are at fault. There has long been a collusion in our society which goes something like this: 'I as a patient don't take responsibility for my health as an independent person, instead I'll have this magical authority figure who will do it all for me so I can remain passive and give all the power to him.' Some patients do play that game and doctors respond to it. So it requires shifts on both sides, on that of the doctor and most certainly on that of the patient.

But that still leaves the second of the two groups of patients, those who for a variety of perfectly valid reasons are not at present in a position to take responsibility for their own health. He saw the solution to that problem as partly a matter of approach:

One of the problems about health education is that it has been tackled almost solely through information exchange, through pamphlets and literature, whereas clearly it seems to me that if you want to change behavioural habits it has to be done through experience, i.e. classes in which there is not just an intellectual exchange of information but giving a person the experience of what it might be like to be relaxed, and that experience which the person has becomes the motivation for continuing.

It's no good saying you ought to do this, or you must do this, or you shouldn't do that, it's in the nature of human beings that they won't respond to that; it's much more to do with creating an atmosphere or a structure in which a human being has a series of choices. It's no good expecting people to change if they haven't got any choices. For some of the population there are no choices so that is the first thing you have to give them. That means tackling questions of housing, of employment, of poverty.

One of the real flaws in what might be called the 'positive health movement' is the danger of moving totally away from the 'government should do more' or 'doctors should do more' to a totally do-it-yourself kind of attitude. To tell someone they have to do it all themselves is just as illogical and unnatural as thinking that doctors have the answers to everything. We live very much in a participatory world – a world which is about living with and among other people, about creating support – and that is to be encouraged. I think it's terribly important not to assume from one's own social class and one's own rather privileged position what other people should or should not do.

I think that what is required and what we are moving into is the knowledge that there are different sets of realities and different sets of truths and that your perception of reality and truth may differ from mine. It is not that you are wrong and I am right but that our realities are influenced by our own past experiences and how we view the world. We have to learn that we have developed a way of seeing the world which is only one way and this means greater tolerance in the way we respond to people's needs. A great flexibility.

This is particularly true when considering the field of complementary medicine. It is clear that orthodox medicine has a place but so do complementary therapies in some cases of patient needs and the problem of trying to bring the two together in the way which is most beneficial to the patient is something we have to learn to tackle.

Dr Tony Perrett is a senior consultant physician at the Royal Cornwall Hospital in Truro in Cornwall. He is fully committed to the National Health Service and does not undertake any private practice.

This is Dr Perrett's view:

The first and most basic need of the National Health Service is that it be properly funded. Funding is now limited to such an extent that one can't even keep up with inflation, the demands of an ageing population or new technology. None of this is being dealt with. Everything is now so pared to the bone that proper

planning can't take place – and when you do plan, then
the plans that you do make are overturned by some
gimmicky political initiative of some sort which diverts
those plans to some other project. This might mean one
year looking after the elderly, next year it might be
the mentally handicapped and another year cervical
screening.

Yet the same basic services have to be maintained and
there is never any extra finance to cope with these new
so-called initiatives. We're now seeing just how these
planning initiatives can backfire – all right, so you launch
a cervical cytology screening campaign, but when the
people come tumbling in, then even if the pathology
services can cope, what happens from the gynaecological
point of view? There's to be no increase in the gynaecological services available to cope with the increased
demand that women are going to want in order to cope
with the abnormal smears. The same will happen with
the Government's so-called initiative on mammography.
These kinds of initiative should be centrally funded.

One could go on and on. Then there are the inevitable
pay rises every year. They have to be found out of the
annual budget and there's no increase in finance to pay
for them. At present the health service just keeps its
head above water but is really providing only a minimal
service. There are always medical advances taking place,
but they must inevitably take place at the expense of
necessary basic services somewhere else. We are all the
time competing across the board for the same amount
of money. The problem besetting the National Health
Service is not its organisation or its management structure, basically it is its underfunding. That is not to say
that there has not been room for economies in some
areas – there has – although I would say that across the
board the use of funds within the NHS has been very
prudent but the economies which can be made now are
negligible. It is really quite unrealistic and parsimonious
of the Government to say that there can be more econo-

mies and that further development within the NHS can be funded by those economies.

If this situation continues, then the basic standard of services is going to fall. That is inevitable. First there is an overall shortage of beds; and, more desperate perhaps, there is an overall shortage of nurses with more and more nurses now leaving the service because of poor conditions and poor pay. Because of the shortage of beds there is a very high turnover of patients, with the result that people can be incompletely looked after. This leads to a lack of job satisfaction for nurses and, indeed, for doctors, and this means the patients get a raw deal; and it is getting worse. It's very hard now to recruit nurses and even harder to retain them and this is one of the most desperate needs at the moment. It's the clinical nurses I'm talking about and unless you are prepared to pay them more and give them better conditions of service then standards of care are going to drop even further.

To keep the standards as they are now a lot of unpaid work goes on, not just by doctors and nurses but by ancillary workers too.

Unless there is proper funding then we are certain to end up with a two-tier health service. Statistically this is a trend which is already becoming apparent as more people start buying health insurance. Presumably this is what the Government would like to see happening.

When you ask where money should go if there were adequate funding then it's hard to know where to start as the needs are right across the board. There certainly needs to be more money spent on primary care, particularly on nursing care and care in the community, but the acute services are very badly served as well. They can hardly keep their heads above water as the demand is too great. So we need more beds, more operating theatres, more consultants, more and better outpatient departments, more and better support services such as laboratories, etc. You see, there is a shortfall in all areas. In addition to what has already been said there also needs to be a major capital programme of replacement hospital building. Half the hospital beds in the country are still in hospital buildings built before the First World War.

This requires a complete change of government attitude, and in all fairness I think all the previous Governments of both political persuasions have under funded the health service. Also, one would like to see a truly independent body governing it which was given its own proper budget so that services could be properly planned, not subjected to frivolous initiatives by various Ministers who are keen to make their mark on the country and leave something behind them for posterity. As I said earlier, senior medical staff and managers spend an awful lot of their time planning for schemes which never take place or if they do are almost immediately replaced by some other, often well-meaning, initiative, by yet another Minister.

It has hardly helped that we have had so many Health Ministers over the last few years and one gets the feeling eventually that like most politicians they are concerned primarily with their own images and their own political futures, rather than with the future of the National Health Service. They really are not interested in long-term planning.

Looking outside the acute services then it is very, very difficult to get proper health care to deprived people. For the most part they tend not to take up what is available mainly because taking any kind of initiative requires effort and this section of the community can become so demoralised that they can no longer see much future even in good health. It takes a lot of indoctrination in health education, and I guess that in the end if you have something like a screening programme then you just have to take it to them. It's the only way to do it. So you need things like mobile screening units as well as an absence of barriers to health care. So you need to ensure that they don't have to pay for the items they need, whether the need is for dental care, spectacles or prescriptions. These have now become real barriers to poor and deprived people getting what they need. Nor are they good at coping with all the forms and regulations which should enable them to get benefit. These facts are well known.

There is no doubt now that there is a very real correlation between unemployment and poverty and deprivation, and between poverty and deprivation and ill health – and not just physical ill health. It leads to a higher proportion of depression, neurotic illness, attempted suicide and actual suicide. These rates are all increasing among unemployed people.

At the end of the day it comes back to the question of political will. The present Government is trying to run the National Health Service like a business, hence all the economies; but, in the end they become self-defeating, for they are economies which deprive people of their essential health care. If you let people live in deprived circumstances, in poor housing and so on, then, as a consequence, you get greater demands for health care, and this is a very expensive way of doing things. An acute bed in a district general hospital these days is costing the country somewhere in the region of £120 a day.

Lastly, people themselves simply do not complain enough about the service they are receiving. I think what they put up with is extraordinary. The obvious example is men awaiting prostrate operations who have catheters put in and who have to put up with that for months at a time or those with arthritic hips who are often in very severe pain but also more or less immobilised and, because of that, demoralised too. Their misery can go on for years. Why they don't complain more, I just don't know.

Occasionally they complain to their Member of Parliament and then he or she will make a fuss and they might get pushed up the waiting lists which, by and large, are a scandal and the present Government initiative of shoving £x-million into the melting pot to reduce waiting lists is absolute codswallop. We're back at the beginning again – for in most places operating time is full, there are still no beds, there are not enough nurses to open new beds even if they are there and the pool of locum surgeons which might be available to do this work is likely to be used up within no time nationwide. So we're back again to underfunding.

AIDS – The Joker in the Pack

AIDS:	Acquired Immune Deficiency Syndrome
Acquired:	This means that this disease is not an inherited or genetic condition. It is now believed that it is caused by a virus.
Immune:	means that this virus attacks a portion of a person's immune system. The immune system is the body's natural protection against infections and disease.
Deficiency:	refers to the defect in the immune system. The portion of this system most affected by the AIDS virus is the T-cells in the blood. T-cells protect against certain infections and possibly some cancers. T-cells are destroyed by the AIDS virus, leaving a person susceptible to these specific infections and cancers.
Syndrome:	means that the acquired immune deficiency is characterised by a group of signs and symptoms.[1]

There are two views as to the effects of AIDS in the NHS. That of the optimists is that it will punch a huge hole in the country's health budget, requiring a massive rethink. The pessimists forecast tens of thousands of cases by the end of the decade with every hospital bed in the country taken up by someone suffering from AIDS or an AIDS-related condition.

When discussing the future of the NHS, AIDS is the joker in the pack. Therefore this chapter has to be taken separately from the rest of the book because we quite simply do not know what is going to happen.

But AIDS and the general thrust of what should constitute a National Health Service do come together quite definitely in the areas of health education and preventive medicine for, put quite simply, we can only talk about AIDS in terms of health education and preventive medicine since there is no cure for it. At the time of writing it is always fatal. Therefore the only way to ensure that you do not die of it is not to get it in the first place – a bleak philosophy.

As we are reeling from a media drench on AIDS it is difficult to imagine that there is anyone now who does not know about it. Yet it does seem that when it comes to funding for research and for treatment and care, as well as for publicity on what it is, how it spreads and how to avoid it, too little was done too late.

We had an excellent opportunity of avoiding what happened in the USA by using their experience, but we did not take it. AIDS first came to general notice in 1981 in the USA where, within a very short space of time, hundreds of cases were diagnosed. In this country the number of cases is now doubling about every nine months.

Nobody is sure where AIDS started but it seems most probable that it originated in central Africa where it is now reaching epidemic proportions in both men and women. The position is so grave in countries like Uganda, where AIDS cases are doubling every six months and where up to 10 per cent of the sexually active population is infected with the virus, that one specialist told the *Guardian* in February 1987: 'I'm not pessimistic but I'm fatalistic. If nothing happens and if we don't beat AIDS, it will be the end of a continent.'

The AIDS virus is known under two different names but in this country we call it Human T-lymphotropic Virus (type 3) or HTLV3. So far as we know at present only about half of those who become infected with the virus go on to develop full-blown AIDS. But blood tests can show that they have been in contact with it because of antibodies found in the blood. They have become antibody positive. We do not yet know why some people do and some people don't.

There is no test specifically designed to diagnose AIDS; it is only when a person develops one of what is known as the 'opportunist infections' that AIDS can be diagnosed. AIDS attacks the immune system of the body, leaving it unable to fight off certain infections and cancers that a healthy immune system could destroy with ease. It is because these infections only occur when the opportunity is right – that is when the immune system is weakened – that they are called opportunist infections. They include a certain kind of pneumonia (pneumocystis carinii) and a cancer called Kaposi's sarcoma (KS).

In spite of the fact that both sexes are equally liable to contract AIDS in Africa, most of those who contracted it in the USA in the early days were homosexual. This led to the spate of headlines about the 'Gay Plague' and a good deal of hypocritical and moralistic publicity on the subject. Indeed, so far in this country, AIDS has appeared predominantly among homosexuals but this is changing. Other groups include haemophiliacs, who caught it through using infected blood products (Factor-8, used to produce blood clotting) before stringent checks were made to ensure that this did not happen; also drug addicts who share needles.

Women have contracted AIDS through sharing dirty needles but also by being infected by bisexual or haemophiliac men. In some cases too women have contracted it through infected blood used in blood transfusions, although the Blood Transfusion Service has now introduced special measures to reduce the likelihood of this happening in the future. There are a few cases, both here and abroad, of babies being born with AIDS, the virus having crossed the placenta from the mother.

Again, as almost everyone will now know, in spite of all the hype AIDS is quite difficult to catch. It cannot be caught by being in the same room with an infected person, by ordinary contact with them, by sharing the same office or public transport. It is transmitted from person to person only through the exchange of body fluids that contain the virus, especially blood and semen. Therefore the most common ways of becoming infected are through sexual intercourse or through sharing dirty needles when injecting into the blood stream, but primarily through the former. Those people unfortunate enough to contract it through infected blood in transfusions or infected Factor-8 did so in the early days of the disease reaching this country – which does not make it any better, for it should never have been allowed to happen, given what we already knew from the United States. It took six years from when it first became apparent as a real threat for the Government to actually take real steps to do something about it.

Throughout 1985, as public anxiety and criticism from the medical profession grew, the Government began to announce a variety of new moves and some extra funding. But the sums of money trumpeted as being newly set aside for AIDS were not what they seemed. To begin with, money for health education on AIDS was to be channelled through the new government committee which, as we have seen, was to replace the Health Education Council. In practice this would mean very little extra funding and other health issues, such as smoking and diet, are likely to suffer as a result.

In fact thirteen specialist health advisory groups will be axed with the disbanding of the Health Education Council, including the Joint Advisory Committee on Nutrition Education (JACNE) and the Environmental Health Advisory Team which helps local authorities to improve food hygiene and establish no-smoking areas in restaurants and offices. Work on all this will go by the board as money is channelled into the fight against AIDS. This is not sensible health education.

The first priority for government money was advertising and this included two television advertisements of doubtful value, one featuring an enormous tombstone and the other an iceberg telling people that safe sex required the use of a condom. (At least it was a breakthrough to allow the word to be mentioned on television at all.) There was also a simple leaflet which was supposed to go out to every household in the country explaining what AIDS was and how to avoid catching it. The latter was not as punchy or clear as that produced earlier by the Health Education Council.

By this time the health professionals were crying out for more funding – for beds, for clinics, for counselling and for the dying who would either need hospice-type treatment or back-up care to enable them to die at home. Again, 'new money' was announced for this, but when the regional health authorities began to look at the small print of the supposed extra funding it was not what it seemed.

For instance, allocations for the three Thames regions announced in February 1987 suggested that the Government was putting £4.4 million of new money into AIDS treatment. But in actual fact that sum included money already allocated for the current financial year and the 'new' money – as the *Observer* pointed out on March 1st – totalled only £1.9 million.

At the time of writing doctors are predicting that at least £20 million will have to be diverted from existing patient care to cope with AIDS by 1988 because of low funding now. Figures provided by the DHSS itself show that the treatment of AIDS in 1988 will cost between £20 million and £30 million for which nothing like such an extra amount of cash is proposed. The planned spending for 1987/8 at the time of writing is actually £12.5 million.

A breakdown of figures shows the position starkly. The Bloomsbury District Health Authority assumed, from government figures, that it would receive extra money for 1987/8 out of the £1.2 million supposedly set aside for the NE Thames Region. Yet, as Dr June Crown, district medical officer for Bloomsbury, told the *Observer*: 'When we looked into it, we discovered that next year's allocation incorporated the £1,045,000 we are already getting for AIDS patient care at the moment. So the extra money for next year will only be £155,000. Because of this the health authority had already had to divert £90,000 from other sources.'

The medical officer for the NE Thames Region, Dr William Kearns, asked how, if £1,045,000 was supposed to be right for 1987, £1.2 million could be right for 1988 when cases were doubling in less than a year. The SE Thames Region is to receive £700,000 in 1988, only £200,000 more than in 1987. They had asked for £4 million. They cannot even cover the costs of £850,000 needed to pay for the provision of heat-treated Factor-8 for haemophiliacs.

But the London Regional Health Authorities are fortunate by comparison with those outside. While the disease has now spread to all fourteen Regional Health Authorities, and to Scotland and Wales, no one else is to receive any extra money whatsoever. In Manchester, for example, doctors are asking for at least £400,000 extra: 36 cases have already been treated, 40–50 are expected to need treatment this year and probably between 90 and 110 next year – at least.

There is equal controversy over the money proposed for research into the disease: £14.5 million to be spent over three years on research into new drugs and vaccines. Some scientists say that this will merely duplicate work already being carried out elsewhere, especially in the USA, and it could be pointless.

The difficulties of coming up with a vaccine which works are immense, not least because it is now apparent that there are at least a dozen strains of the AIDS virus and that the virus itself appears to be able to mutate.

Calls for drug addicts to be allowed to draw free disposable needles also fell on deaf ears for a long time. The problem first became really apparent in Edinburgh, which has a growing population of hard-drug users and where about 50 per cent of them are now thought to carry the AIDS antibody. Towards the end of 1986 Secretary of State Norman Fowler visited Holland where he saw free needles being handed out on a regular circuit. Not only was this helping to prevent the spread of AIDS, it was also making it easier for the authorities to discover just how many drug users there were, to offer them help to come off drugs and, in some cases, to pick up the pushers.

(In passing, if it seems that too little was done too late about AIDS then the situation is even more shameful regarding drug abuse. Again there was ample warning of a pending epidemic, again the response was minimal consisting mainly of an advertising campaign considered, by virtually everyone outside the DHSS and the advertising agency that produced it, to be absolutely useless and an appearance on television of Mrs Thatcher in a Customs Office at Heathrow looking the camera in the eye and saying the Government was out to 'get' drug traffickers. At the time of writing a number of pilot schemes are in operation in selected parts of the country where drug addicts can get free disposable needles in exchange for their old ones.)

Following the high profile given to AIDS by the media – all four television channels ran whole series of programmes on the subject in early 1987 – there has been enormous pressure for blood tests. Again there is insufficient laboratory or clinic capacity to deal with the demand, let alone enough counsellors to help those who find that they are, in fact, antibody positive.

Dr B. is a senior consultant in genito-urinary medicine in an area outside London and he is responsible for looking after AIDS patients. He does not want publicity by name for the simple reason that he does not want to prevent those people in his own area who suspect they might have the disease from coming forward because they realise how few resources are at his disposal. He himself is currently running several regular clinics free of charge as no money is available. This is what he says:

Money – are we going to get it? At present there seems precious little chance that we will get anything like what we want. This will be the ongoing problem in this and other parts of the country. Obviously you recognise that, with only finite resources and an infinite demand, there will never be sufficient funding within the NHS to cover everything. We all know this. But AIDS is a special case.

What has been done within our own district is typical. The District Health Authority has set up an AIDS Action Group made up of doctors with different specialities. It meets only two or three times a year, it has no teeth, it has no budget, it has no resources. It is simply an advisory committee. It can make recommendations, and indeed has done so. By setting up such a group it can be made to look as if, on the surface, things are moving along quite nicely. In reality, if you ask for anything, then your request is referred to the AIDS Action Group. As they have no money at all this produces something of a predicament so they refer the request back to the District Health Authority which, in turn, refers it to the unit general manager who then refers it to the district general hospital which refers it back – you've guessed it – to the AIDS Action Group!

Nationally all the money – and that in itself is totally insufficient – has been allocated to London. Nothing at all has come out to the districts. I've even checked with the Regional Health Authority and they have not received a single penny for AIDS and they tell me they are not likely to.

Of course the position in London is acute but the

disease is not going to stay in London, something which I pointed out personally to Mr Fowler when I met him recently. Many of the people attending London hospitals are, in fact, from the provinces – a point which does not appear to have been truly considered. The majority of my own patients have become infected in London and then have found that their friends have deserted them, they do not want to know about them, their employers do not want to employ them and all that is left is for them to come home to die. They can only do that if their families are willing to have them back and care for them. But the money that has gone to London to treat patients does not come back with him/her when he or she returns home – which is what I told Mr Fowler. But it does not seem to have made any difference. The money is still allocated to London.

I have had two or three patients who had to draw money from the DHSS for the train fares to go to London for treatment and the DHSS have got annoyed and asked me why they cannot be treated here and I have had to explain that we have not got the facilities. The Government must be made to realise that all those being treated in London are not necessarily from London and so the money needs to come back to the regions. Yet no one is getting anything anywhere – and that is the brutal truth of the present financial position.

It is all very well for the Government to look at the regions and see that compared to London they are comparatively safe with regard to AIDS, outside the big cities. But when you see how quickly this disease spread from Africa and the USA to London then it is quite obvious that it will rapidly spread elsewhere – we cannot build Berlin Walls around our communities to keep it out.

Some authorities like to say they have no cases at all unless they have actual full blown AIDS cases but most must now have a number of patients who are antibody positive and beginning to show that they are in danger of developing AIDS. I have a number of these under my care already and I already know that I will have more as I have been told by other consultants of their patients who will be coming into my area and therefore into my care.

I know of one who will be coming to me from America. He's gay, he's single and he wants to spend some time with his family in the place of his birth before he faces up to going back there and dying in the States.

He told of a number of other tragic cases, cases, he says, which are all too typical of what is happening today.

They come to me and I feel helpless. I have to tell them there is no vaccine, there is no cure. I cannot offer them anything. One patient asked me if there was anywhere he might go so I told him there were only two places doing real research into AIDS, France and America. So we discussed it and I told him I did not think he should go to France as he did not have the language. So he sold his house and went to America and got a job so he could pay for his treatment – he did not tell his employers he had AIDS – until his condition suddenly worsened and he went from bad to worse until eventually he was given only three months to live. So he rang his boyfriend over here and told him he did not want to come back and die because all his relations would be embarrassed, so he made a suicide pact with his friend which he carried out.

Patients to whom he had had to break the news that they have AIDS have actually begged him, on some occasions, to give them something, poison, sleeping pills, so they can end their lives, pleading that they would do it outside the hospital and not involve him. Needless to say he has not done so.

This is not uncommon. When I heard of the death of the patient mentioned earlier I set up the first small.

research unit to look into how best to care for AIDS patients. The Terrence Higgins Trust, Body Positive and Lighthouse, etc., look after homosexuals but it is now no longer just homosexuals who are getting AIDS. The reason no one had done any proper research into the best way of caring for AIDS sufferers is because the government did nothing for so long. AIDS was the 'gay plague', a disease of homosexuals, so the government was little concerned. The attitude seemed to be 'let them die'. Then it began to spread – to haemophiliacs who had had untreated Factor-8, their wives, drug addicts, even to babies.

I had another patient who chose to die outside his own area and go to London as he did not want to inconvenience or embarrass his family. I used to try and go and see him when I was up there and it was terribly sad, for he was very unhappy dying up there on his own – these are real tragedies.

But you can see why I don't want to spell out the position clearly within my own area and say that I haven't got the staff or the resources or even a counsellor to talk to someone after he or she has had a positive antibody result from his/her blood.

AIDS poses a number of moral problems, and not just those touted around by people with fundamentalist religious views. Because of the extreme reaction to those not only with AIDS but even with positive antibodies, there is a danger of the disease being driven even more underground and therefore spreading more easily. Some of the organisations set up to help homosexuals appear to be advising their clients to keep quiet if they are antibody positive, as frankness can prevent them from getting a mortgage, a job or even a college place. I came across one instance of a boy who was asked to leave his college merely because he was known to be a homosexual although he had willingly undergone a blood test and been found to be negative.

AIDS is not a notifiable disease, and this poses a real moral dilemma to doctors, as Dr B. points out:

Ultimately most of the AIDS patients will go back to their communities to live. I think if parents are looking after them, then they should be told the truth even if that person is gay and has never admitted to his family that he is a homosexual. But by not telling your parents you could, in some circumstances, pose a risk to them because they do not know. This can happen in the case where, say, an AIDS patient starts bleeding. There was one such case where this happened and his mother used her fingers to put pressure on the bleeding point but she herself had an open cut on her hand. Perhaps then an ambulance is called and the ambulancemen have to put a dressing on a wound and they are at risk too.

There is a great need for counselling, for the setting up of a proper counselling service. Not just to explain and help the patient after he has been told he is antibody positive or even has AIDS, but to help him to explain the position to his parents and to encourage him to do so. I myself went on an AIDS counselling course to find out what it entailed and I then sent a couple of members of my staff. It was helpful, but we need full-time counsellors.

I think perhaps AIDS should become a notifiable disease for this reason. As a doctor I am bound by confidentiality. I cannot even tell a patient's GP that he has AIDS if he has been referred to me by another doctor and he does not want his own doctor to know. So if he falls ill his doctor does not know how to treat him.

Even worse is the kind of case I have come across where I have treated a patient and have found he has AIDS. Either he is bisexual and has gone with men or he has contracted it from a prostitute. I know that he is still having a sexual relationship with his wife and he has not told her that he goes with men or with call girls, but

I cannot tell her because of the legal problems. She realises there is something wrong because of the fuss, but if she asks me directly then I cannot say 'Yes, your husband has AIDS', I can only tell her he is having some kind of a problem and then try and persuade him to tell her himself, often without success.

My position is intolerable. I know he is virus positive, I know he is sleeping with his wife and not telling her, but I am bound not to tell her because it is not a notifiable disease. But what is his position – should he not, in the extreme, be tried for murder or manslaughter, for he is almost certainly infecting and killing his innocent wife?

Dr B., like others in the medical profession, sees the need for counsellors as a prime one – for the AIDS sufferers, their relatives and friends and, indeed, for the public at large. Dr P. Donaghy, Specialist in Community Medicine for Cornwall, feels that it is extremely important that as many nurses and health service personnel as possible are educated about AIDS and its effect: 'They act as good carriers of information – spreading it almost like the way Christianity spread in the early days, from door to door and over the garden fence.'

The gay community, against whom there is the most prejudice, get what help and support they can mainly from the specialist groups like the Terrence Higgins Trust or from their own self-help groups. For instance one group of antibody-positive gay men in Plymouth set up their own support group when they found there was none available, assisted by the newly recruited AIDS counsellor in the city. Some admit that they are no longer acknowledging that they have the disease or even seeking to know whether or not they are antibody positive because of what this might mean. They have to cope not only with coming to terms with the notion that they may well develop a terminal disease, but also with knowing how they are likely to be treated by the rest of the community, no matter how often

people are told that AIDS is difficult to catch outside a
sexual relationship with an infected person.

Nurses too require special counselling, for although they
are all too often faced with those who are terminally ill, in
this case they are dealing with predominantly younger
people, much their own age, for whom there is no hope
at all.

Informed medical opinion sees the need for flexibility of
funds for treating patients with full-blown AIDS. There
will have to be acute beds for those who need treatment
for AIDS-related diseases such as pneumocystis carinii and
Kaposi's sarcoma; there may well have to be a number of
isolation beds for those patients who are excreting a lot of
bodily fluid. Finally, for those near to death, there will
have to be either hospice care or assistance for those who
are looking after them at home and that certainly means
extra resources, not least for the community nursing service.

The hysteria will also have to be taken out of the subject,
especially in the popular media. I have been told on
good authority that some popular newspapers are offering
'moles' in hospitals several hundred pounds for details of
any AIDS patients being treated there, with a bonus if the
person is likely to be well known.

There must be no let-up in the campaign for 'safe sex'.
It is highly unlikely that there will be a mass return to
Victorian values and if there were it would hardly help –
the massive number of sufferers from venereal disease in
the nineteenth century showed that, whatever might be
said about morals in public, private practice was manifestly
different. So there will be a continuing need for the kind
of publicity some find distasteful – pointing out that rectal
sex is the most dangerous of all, that condoms should be
used unless couples are sure that neither could be a carrier
of the AIDS virus and that dirty drug needles can also
spread it. Dentists, acupuncturists, tattooists and ear
piercers will have to ensure that everything they use is
properly sterilised – although it is unlikely that there are

any dentists left who do not take such a precaution. Blood donors will obviously need to be cleared and there must be no repetition of what happened with Factor-8. Had the necessary funds been found for the expanding of the transfusion service, long sought, then we would never have needed to buy in blood from the United States with its associated risk.

Meanwhile research goes on. Scientists at the Institut Pasteur in France, among others, are working on AIDS vaccines. The major problem with the AIDS virus is that it invades and kills the very cells that are supposed to protect us from infection: so that when a person contracts an infection and the cells which are supposed to defend him become active, so does the AIDS virus. So a vaccine will have to do two things – generate an immune reaction to every type of AIDS virus, and also prompt the system to generate antibodies capable of destroying the cells in which the virus lurks.

Research and development is also going on into drugs to be used for AIDS patients. Most of those currently being tested are designed to stop the virus reproducing after it has entered the cell. A prototype drug developed in France, known as HPA 23 was used in one case and appeared to halt the march of the disease. Within days of a report on this being made public the Institut Pasteur was inundated with calls from people from all over the world asking for it, while dozens of Americans turned up on its doorstep pleading for treatment – among them the film star Rock Hudson who died shortly afterwards. While there was some limited success with HPA 23 in halting the disease for a short time, nobody thinks it is the answer to the problem.

Recently the Government has licensed AZT, a drug developed by the Wellcome Foundation which appears to halt the disease by interfering with the DNA chain. It has already been used on a limited number of AIDS patients, both orally and intravenously, at the US National Institute of Health. It was tried on nineteen patients, fifteen of whom showed an increase in the number of the vital T-cells which are the front line of the body's immune system. It appeared to give those patients a remission in their illness, actually allowing the most emaciated to gain weight.

But there are a number of highly unpleasant side-effects in many cases, and some doctors have reservations about how widely the drug should be available so early in its development. Even if the side-effects can be adequately dealt with, this drug, at best, can at present only offer the hope of remission.

So the outlook remains bleak. To try and see exactly what the position will be, even by the end of the decade, would take, says Community Medicine Consultant Dr Donaghy, 'a crystal ball'. At first it looked as if only 10 per cent of those with positive antibodies would develop the disease. Now it seems as if about 50 per cent will go on to develop it and it might well be that everyone who is antibody positive will eventually develop full-blown AIDS. It is still too early to tell. The incubation period of the disease is also continually being lengthened. At the time of writing there have been 878 AIDS patients of whom 490 have died.

Just why the virus has arrived on the scene now nobody knows. There have been a number of theories, from it being an escape from somebody's germ warfare pro-gramme – either Russian or American; that it has been around all the time but that pollution, particularly in the form of the pesticides designed to attack DNA, has lowered everyone's immune threshold and thus made us susceptible to a disease we would never normally have caught; that it is, indeed, a new disease which has appeared spontaneously.

Finally, to quote Dr Donaghy again:

AIDS was a wonderful disease for all kinds of people, most especially for the popular press. It had everything – it was a gay plague, a disease of moral degenerates, those who caught it deserved all they got. It provided them with marvellous headlines so it was a fine disease for the media.

Then, from the research point of view, it was also a splendid disease and there's probably never been one into which more effort has been poured to try and find a vaccine and a cure. Then there are those who write long articles in the learned journals, it's been great for them too. It was fairly good for the authorities as well, because for a long time they did not take it seriously because it came over from America, from the gay community, and, whatever we might say about our supposed national tolerance, there is inherent prejudice against homosexuals.

But for the unfortunate person who is actually suffering from it, it is an absolute tragedy, a real tragedy for it is more certainly terminal than any of the known cancers – even the very worst.

12

Summing It Up

So is there a future for the National Health Service? It is now apparent that had we been setting it up today then, irrespective of political ideology, it would have been done differently. As it is, it is with us warts and all, firmly based on the ideas of forty years ago, although it is under increased threat.

While it is also obvious that the demands made upon it are infinite, its funding will always be limited to what it is felt the country can afford – although how high the NHS figures on the list of a government's priorities is very much a matter of political ideology.

What certainly seems to have happened is that although there have been successive changes in the bureaucracy, these have not helped those working at the sharp end. There is no more fat left to cut. As the new managers are now discovering, the ability to run a supermarket chain or command a battleship successfully does not necessarily equip you to run part of what should be a caring organisation where the needs of people should come first.

Also, as we have seen, loud proclamations about the amount of money spent on health, which bear little or no resemblance to what is being experienced in real life, cut little ice. Possibly if those who had so much to say about how splendid everything is actually had to use the service themselves, things might be very different.

Two stories picked almost at random from the *Guardian*

of March 30th bear this out. One, headlined 'Hospitals Face Cuts in Doctors for Efficiency' says that senior health service managers in Rochdale have decided that their doctors are treating more patients than they can 'afford', as they have been told by central government to 'save' £4.8 million in three years. The reaction of the consultants and doctors involved is one of rage.

In a letter released to the press they say the cuts are totally unacceptable: 'Major cuts last year have reduced the level of services we can provide for patients to potentially dangerous levels. Further cuts would mean that patients from Rochdale might have to seek care in other districts who are themselves hard-pressed.' Dr Christopher Davidson, a consultant physician, described the plans as 'nonsense', pointing out that the authority wants to get rid of whole teams of medical staff from the consultants downwards.

Rochdale has, in fact, done everything central government demanded. It has treated more patients in few beds faster and it has reduced its waiting lists. Other districts within the North West Region are also, at the time of writing, in grave difficulties because the whole health strategy for the region was based on a forecast that the numbers needing acute treatment would fall! Mr Denis Allison, the region's general manager, admitted to the *Guardian* that he had got his forecasts wrong 'but the budgets have already been set'.

The second story, headed 'Abusing the Civil Service', by Richard Norton-Taylor analyses how, increasingly, civil servants are becoming servants not of the Government in the widest sense but of the Party in power – and there is a distinction.

Amongst the instances he gives is how at the very time civil servants from the DHSS were trying to block the publication of the Report 'The Health Divide – Inequalities in Health in the 1980s', published by the Health Education Council, other civil servants at the Welsh Office were busily distributing free glossy leaflets to hospitals and doctors'

surgeries which concluded with the bold statement that
'the Health Service in Wales has never been in better
shape'. At least they will provide reading matter for those
waiting endlessly in queues for treatment.

The two stories sum up the situation as it is. We have
an underfunded acute service desperately trying to achieve
even a basic standard of care, while outside it the general
health of the growing numbers of the deprived deteriorates
rapidly.

The Black Report, 'Heartbeat Wales', the Sheffield
Survey, 'The Health Divide' – how many more Reports
do we need to tell us that the situation is scandalous? That
our housing stock is so depleted and so poor that tens of
thousands of people are living in squalor or have no homes
at all? That mass unemployment produces severe physical
and mental health problems? That low wages bring in their
wake, among other things, poor nutrition? That tens of
thousands of children are living in real poverty without
even the old safety net of nutritionally balanced cheap
school meals?

And it *is* a scandal. The hypocrisy of those who go
abroad and purport to tell other countries how to organise
themselves and improve their system beggars belief. If
proof is needed that we have become a seedy, divided
nation, divided between those who grab all they can and
turn a blind eye to need and those who are now too
defeated and apathetic to do anything about it, then it is
there in the nation's health.

However, let us assume that one day the realisation
will dawn that a healthy population would make fewer
demands on the National Health Service, not more; then,
hopefully, some of the ideas put forward in this book might
point the way.

We need an energetic commitment to the concept of
primary care, with special emphasis being given to services
for those in the most deprived areas. There should be an
immediate reinstatement of a fully independent Health
Education Council – we need more independent health

education, not less. We have to find a way of getting across the message that we are, in part, responsible for our own health – we do not need to smoke, drink too much or eat rubbish food; but, as both Dr Patrick Pietroni and Jean Robinson rightly point out, those who are most in need of the message are those least likely to be able to cope with it. If you are living in bad conditions, unemployed and deprived, then your main business is merely to survive and health education, however worthy, seems pretty irrelevant.

Certainly the example of the Peckham Experiment should be examined closely, the idea of positive health, a 'well persons' centre', offering a whole range of leisure and social activities, along with health check-ups and medical care. While a modern version of such a scheme nationwide might be expensive to set up initially, it seems almost certain that in the long run it would save money for it would raise the level of general health.

There is certainly a need, too, for a very different philosophy for the National Health Service. It desperately needs more democracy. It would surely be possible to find a way whereby those who sit on Regional and District Health Authorities and Community Health Councils are elected, not appointed? So that they (including myself) are accountable to those whom they are supposed to serve.

It also needs a fairly massive rethink by the medical profession so that it is no longer so doctor-dominated, something which is widely recognised by the more progressive members of what is one of the country's most conservative professions.

For one of the most major shifts that those who set up the NHS could not have foreseen is the growing part now played by the patient, consumer, client, whatever term suits best. Outside the ranks of the unfortunate beaten-down deprived section of the population are those who either are not wealthy enough to afford private care or who are, anyway, opposed to the idea, but all of whom see themselves as being under-represented when decisions are made about the NHS, and who should be encouraged to have far more input.

In fact, the Peckham Experiment, itself set in a deprived area dealing with the very kind of people now featuring in the Black Report, found by trial and error that it became successful only when it had truly involved everyone in the neighbourhood in their own health care. It is worth quoting what Dr Pearse said about it once again:

> The doctors learned from the people. Our failures . . . taught us something very significant. Individuals from infants to old people resent or fail to show any interest in anything initially presented to them through discipline, regulation or instruction, which is another aspect of authority . . . We now proceed by providing an environment rich in instruments for action – that is, giving a chance to do things. Slowly but surely these chances are seized upon and used as opportunities for development of inherent capacity. The instruments of action have one common characteristic – they must speak for themselves. The voice of the salesman or the teacher frightens the potential users.

The future of the National Health Service should be safe in *our* hands. It is too important to be left to government for, at the end of the day – whose health is it anyway?

13

Author's Note

While the first edition of this book was in preparation, the General Election took place. Health certainly featured on the agenda and the author, who was closely involved in providing information on the NHS, was inundated with what can only be termed horror stories. In fact it was the health issue, on what became known as 'Black Wednesday' that alone rattled the Conservative Party in its almost unimpeded progress to victory and which resulted in the Prime Minister making her famous statement about wanting to choose the doctor and hospital of her choice, at her time and when she wanted to.

About a month after the book was published the health issue literally blew up. It was as if a lid had been kept down on a boiling pot for too long and for months and months (as at the time of writing) we have seen nurses, doctors and other health workers on the picket lines, taking part in demonstrations, some nurses on strike and acres of coverage of what is happening in the NHS. All the health workers are making one thing plain – although some of the disputes are over low pay it is the NHS, not personal gain, which is uppermost in their minds.

Along with the action has come a positive spate of reports showing that the NHS, in spite of what the Government has said and continues to say, has been grossly underfunded for years. Before looking at the figures, we can take just a brief look at some of the instances which

have come to light during the last six months – a few out
of hundreds.

These are in no order of importance but give the general
picture of what has been happening. The Birmingham
Children's Hospital hit the headlines with the news that
there is a waiting list of nearly a hundred for children
needing major heart surgery owing to a shortage of specialist nurses. Many babies and toddlers have had to be turned
away several times when their operations have been due
and three babies died, two following surgery and one while
waiting. In the case of the two who died, doctors do not
consider that the wait was responsible – both cases were
far more complex than newspaper reports would lead one
to believe – but in two cases frantic parents actually took
the health authority to court in their desperate fight to get
their children the treatment they so badly needed. Both
cases were thrown out, the judgement being that it was
inappropriate for courts to give ruling on what has to be a
medical condition. Certainly the parents of the children
who died, and who saw them deteriorating in front of their
eyes, feel they would have had a better chance had their
operations been carried out when they were supposed to
have been.

In a search for cuts, Cornwall Health Authority closed
a small general hospital in Falmouth, a maternity hospital
in Penzance and did away with its hospital car service. The
closures first 'temporary', soon permanent, meant that
pregnant women from the western end of the county now
have to travel anything up to 40 miles to get to hospital.
When the MP for Truro, Matthew Taylor, asked junior
Health Minister Edwina Currie how they would manage
such distances in an area where there is much poverty and
low car ownership, poor public transport facilities and now
no hospital car service she replied 'let them take taxis'.

In March 1988 a pregnant woman was turned away from
25 hospitals while on the point of giving birth to premature
twins. She went into labour at the Royal Berkshire Hospital in Reading but there were no intensive care cots

available. The special baby unit already had 18 babies in it and there was only funding anyway for two intensive care cots. Finally, while still in labour, she was driven 80 miles to the Royal Sussex and County Hospital in Brighton where she gave birth to the twins, one of whom died three hours later. She had, she said, nothing but praise for the staff who tried so hard to help her.

Also in March, St Thomas's, London's famous teaching hospital, announced it would have to close 200 urgently needed beds, cancelling 2,100 admissions and 60,000 out-patient attendances, while shedding more than 300 staff as its health authority was over-budget. In the same week the hospital which considers itself to be the country's most efficient, Scarborough Hospital which became top of the Department of Health's performance league, announced that patients were likely to be in peril through such efficiency. Doctors there have said that while the hospital now treats patients more swiftly and in fewer beds than any hospital in England this has been at the expense of worse care, more complaints from patients and low morale among overworked doctors, nurses and consultants.

A round-up of cuts in their own districts by Community Health Councils produced scores of examples of which the following are only a few: temporary reductions of maternity beds and reduction in ambulance car service in Salisbury; closure of 25-bed orthopaedic ward (to be re-opened with 13 beds), permanent closure of geriatric ward and restricted access to physiotherapy in Macclesfield; weekend closure of wards at an eye hospital and two-week closure of a day abortion unit in Liverpool; amalgamation of two children's wards, closure of a surgical ward, postponed opening of a drug misuse service in Bristol; £600,000 to be 'saved' by late paying of bills in the Prime Minister's own health authority district, Barnet; knee and hip replacement operations suspended until the end of the financial year in Clywd; a geriatric ward opened by the Queen Mother in 1987, closed in 1988 through lack of resources in Gwent; closure of five dental clinics and

reduction of sessions at three others, closure of eleven
family planning centres and reduction of services in four
others in Hounslow and Spelthorne . . . and many, many
more.

On 16 December 1987 Health Minister Tony Newton
announced that there would be an extra £75 million available for current expenditure by health authorities in England and Wales but, as the National Association of Health
Authorities pointed out, there was a shortage of £95 million
in the funding of pay and price inflation alone, in 1987–8.
This amount would nowhere near tackle the overall problem and the announcement did nothing to stem the tide of
concern about the health of the service. A further outcry
provoked the Prime Minister to announce on television
that she would be heading a review of the NHS, in which
the private sector will play a far larger role.

In an unprecedented move the heads of the Royal Colleges spoke out on the issue, pleading for more funding
and on 7 December 1987 they issued a statement saying
that the NHS had 'almost reached breaking point'. A
survey undertaken by the Royal College of Surgeons
showed that out of the 151 health authorities which had
responded, 90 had closed surgical beds and a third had
operating theatres which had had to be taken out of use.
On 4 January 1988 the British Medical Association called
for an extra £1.5 billion for the NHS immediately and that
the NHS should continue as an essentially tax funded
service.

On 1 March 1988 the Tory-dominated House of Commons Social Services Committee published its report and
recommendations into NHS funding and called for an
immediate £95 million to make up the shortfall in the
funding of pay and price inflation in the current financial
year; an additional sum of at least £1 billion over two years
for specific developments; a commitment to fund all pay
awards in full and a guarantee to fund a 2 per cent rise
in services in 1988/9. This was the result of weeks of
intensive taking and hearing of evidence. Its findings were

rejected by the Health and Social Services Secretary, John Moore (who had just spent several weeks in a private hospital) and Health Secretary, Tony Newton literally within hours of its publication. At the time of writing an angry committee is prepared to fight it out.

Finally during the first week in March the respected and independent King's Fund published its own report on the state of the NHS and made a number of recommendations which included looking at new ways of rationing services and a greater role for the private sector. But, having said this, it also states that whatever other initiatives are pursued, the case for increased public spending in 1988/9 is overwhelming. Pay awards should be met in full and an extra £400 million made available immediately to compensate for increases in demands on the NHS that have taken place during the 1980s.

Ideas being bandied about in advance of the findings of the Prime Minister's review are hotel charges for patients in hospital, a two-tier system of private and public health insurance and (Mrs Edwina Currie) people having operations performed privately rather than taking holidays or decorating their homes.

On the run up to the Budget on 15 March poll after poll showed that a massive majority of people preferred any spare money – and apparently we are awash with it – spent on the NHS rather than tax cuts. The Labour campaign calling for Budget Day to be made NHS Day was ineffective, too little and too late and the SDP and Liberals were still locked in internecine warfare.

The result was apparent when the Budget was announced. There was to be no extra funding for the NHS. Instead tax was cut by two pence in the pound and the higher bands of tax slashed to 40 per cent – arguably the biggest hand-out to the rich since the Dissolution of the Monasteries.

Notes

Chapter 2: The Jewel in the Crown

1 'Life before the National Health Service'.
2 'Antiquated Hospitals', *Community Health News*, December 1984.
3 'Priorities for Health and Personal Social Services in England – A Consultative Document', DHSS, 1976.
4 'Prevention and Health; Everybody's Business – A Consultative Document prepared jointly by the Health Departments of Great Britain and Northern Ireland', DHSS, 1976.
5 'Sharing Resources for Health in England and Wales – Report of the Resource Allocation Working Party, DHSS, 1976'. (The RAWP Report.)
6 *Report of the Royal Commission on the National Health Service 1979*, (HMSO, Command Paper 7615).
7 'Patients First: Consultative Paper on the Structure and Management of the NHS in England and Wales', DHSS, 1979.
8 'National Health Service Management – An Inquiry' (The 'Griffiths Report'), DHSS, 1983.

Chapter 3: Safe in Their Hands?

1 'Hitting the Skids – A Catalogue of Cuts in London', London Health Emergency, 1987.
2 Information from Cornwall Community Health Council.
3 *Ibid*.
4 'The Babies Who Need Not Die', the *Guardian*, 7.7.1986.
5 'Consumer View – Patients' Choice', National Consumer Council, 1986.
6 'A Review of Outpatients' Departments from Community Health Council Surveys', Association of Community Health Councils, 1986.

Chapter 4: Lies, Damned Lies, and Statistics . . .
1 'U.K. Health All Spent Out', *Health Services Journal*, May 22nd, 1986.
2 'Public Expenditure in the NHS: Recent Trends and Future Problems', Institute of Health Service Management, British Medical Association and Royal College of Nursing, 1986.
3 *Ibid.*
4 'Life Before the NHS', Health Service Committee, TUC East Midlands Region, 1986.
5 *Guardian*, August 12th, 1986.
6 'Health Expenditure in the UK', Office of Health Economics, 1986.
7 Coventry Community Health Council figures.
8 *Guardian*, August 12th, 1986.
9 'Finding a Doctor in Bloomsbury', Bloomsbury Community Health Council, 1982.
10 *Hospital Doctor*, January 13th, 1986; Hansard, January 13th, 1986.
11 'Report to Special Conference of Representatives of Local Medical Committees', November 13th, 1986.
12 DHSS leaflets, January 17th, 1979 and July 1986.

Chapter 5: Be Poor and Die Young
1 *British Medical Journal*, December 15th, 1984.
2 *Royal College of GPs Journal*, November 1985; *British Medical Journal*, November 9th, 1985.
3 Figures given by Low Pay Unit to author.
4 Labour Research Department document, December 1986.
5 *Ibid.*
6 *British Medical Journal*, August 9th, 1986.
7 'The Big Kill – Death from Smoking', Health Education Council, 1985.
8 *Ibid.*
9 'That's the Limit', Health Education Council, 1986.
10 'A Guide to Healthy Eating', HEC, 1986.
11 *Community Health News*, December 1985; Association of Community Health Councils.
12 See *The Price of Freedom* by the author (New English Library, 1985).
13 'Pesticide Residues in Food', London Food Commission, 1986.

Chapter 6: 'I Wouldn't Start from Here . . .'
1 'Public Expenditure in the NHS – Recent Trends and Future Problems'.
2 Information from Cornwall Community Health Council.
3 *Community Health News*; ACHCEW, February 1986.

Chapter 7: The Sharp End
1 'Heartbeat Wales', the Welsh Office, 1987.
2 'Neighbourhood Nursing – A Focus for Care', DHSS, 1986.
3 Information from 'Nurses Do It Better' workshop at AGM of ACHCEW, 1986.

Chapter 8: Positive Health
1 'National Guidelines for Well Women's Clinics', ACHCEW, 1981.
2 'Subjective Measures of Quality of Life in Britain 1971–1975', Polytechnic of North London, 1976.

Chapter 9: The Peckham Experiment
1 The basic information in this chapter was taken from the following sources:
   Innes Pearse and Liz D. Crocker, *The Peckham Experiment* (Allen & Unwin, 1943).
   'An Experiment Interrupted', *Lancet*, February 22nd, 1986.
   Colin Ward, 'Peckham Revived', *New Society*, December 20th, 1987.
   Michael Young, 'The Peckham Experiment', *Self Health*, No.11, 1986.
   'A Reminder and a Memorial', *British Medical Journal*, March 22nd, 1980.
   Joe Elkes, 'Self Regulation and Behavioural Medicine', *Psychiatric Annals*, February 2nd, 1981.
   'Positive Prospects for Health', report of a public meeting held by Pioneer Health Centre Ltd, December 1985.

Chapter 11: AIDS – The Joker in the Pack
1 Linda Moxey and Gayling Gee, 'The AIDS Medical Guide', San Francisco Department of Public Health, 1986.

Information for this chapter has also been taken from the following: 'AIDS and the Community – A Preventive Strategy', Manchester Community Health Councils, 1986.
Terrence Higgins Trust:
'AIDS – the Facts', 1985.
'HTLV–3 To Test or Not to Test', 1985.
'AIDS – You and the Antibody Test', 1985.
'AIDS – Medical Briefing', 1986.
'AIDS – What Everybody Needs to Know', HEC, 1986.
'DRUGS and AIDS' and 'WOMEN and AIDS', by Linda Moxey and Gayling Gee, San Francisco, Department of Public Health, 1986.
'The Epidemiology of AIDS', *Science*, Vol. 229, September 27th, 1985.
'Acquired Immune Deficiency Syndrome HTLV3/LAV – The Causal Agent and Modes of Transmission', *Health Trends, 1986*, Vol. 8, WHO.
'AIDS – Act Now, Don't Pay Later', *British Medical Journal*, August 9th, 1986.
'AIDS – Submission to House of Commons Select Committee', ACHCEW, 1986.
'The Causative Agent of AIDS – Revised Guidelines', Advisory Committee on Dangerous Pathogens, June 1986.
'AIDS – Questions and Answers', by Dr V. G. Daniels, Cambridge Medical Books, 1986.
'More Government Money for AIDS', DHSS Circular 85/264 September 26th, 1985.
'AIDS Extras 1 and 2', *Guardian*, November 5th and November 6th, 1985.
'AIDS in Africa' series, Peter Murtagh, *Guardian*, February 3rd and February 5th, 1986.

Select Bibliography

General Reports by Various Professional Bodies

'A Charter for Action – Health for All by the Year 2000', Faculty of Community Medicine, 1986.

'Action on Outpatient Services – A Time to Move', Institute of Health Service Management, 1986.

'A Manifesto for Nursing Health', Royal College of Nursing, 1986.

'Better Management, Better Health', NHS Training Authority, 1986.

'Health Expenditure in the UK', Office of Health Economics, 1986.

Alan Maynard and Nick Bosanquet, 'Public Expenditure on the National Health Service – Recent Trends and Future Problems', Institute of Health Service Management, British Medical Association and Royal College of Nursing, 1986.

'Planned Health Services for Inner London', King's Fund, 1986.

'Report to Special Conference of Representatives of Local Medical Committees', General Medical Services Committee, November 13th, 1986.

'The Vision – Proposals for the Future of the Maternity Services', Association of Radical Midwives, 1986.

'Your Local Ombudsman', Commission for Local Administration in England and Wales, 1985.

Reports from political parties, trades unions, etc.

'Hitting the Skids – A Catalogue of NHS Cuts in London', London Health Emergency, 1987.

'Hospital Admission Cuts', the Labour Party, January 1987.

'Life Before the National Health Service', Health Services Committee, TUC East Midlands Region, 1986.

'The Best of Health – Charter for the Family Health Service', Labour Party, 1987.

Reports from Consumer Bodies, etc.

Association of Community Health Councils for England and Wales, 'Evidence to the Select Committee of the House of Commons', *Primary Care Review*, 1986.

'A Stake in Planning: Report of the National Council of Voluntary Organisations', Joint Planning Working Group, 1986.

Chairman's Report, Association for the Victims of Medical Accidents, 1986.

'Consumer Voice – Patients' Choice', National Consumer Council, 1986.

'Digest of Cuts and Closures', Greater London Association of Community Health Councils, 1986.

'Food for All?', London Food Commission, 1986.

'Finding a Doctor in Bloomsbury', Bloomsbury Community Health Council, 1982.

'Food Facts', London Food Commission, 1986.

Christine Hogg, 'Community Health Councils – Their Role and Structure', ACHCEW, 1986.

Christine Hogg, 'The Public and the NHS', ACHCEW, 1986.

National Consumer Council Annual Report, 1985/6.

National Consumer Council Response to the Select Committee of the House of Commons' Primary Care Committee.

'National Guidelines for Well Women's Clinics', ACHCEW, 1981.

'Paper on an NHS "No Fault Compensation" Scheme', South Gwent Community Health Council, 1987.

'Patients' Rights', National Consumer Council and ACHCEW, 1986.

'Pesticide Residues in Food', London Food Commission, 1986.

'Subject Access to Personal Information', ACHCEW, 1986.

'The Patients' Charter – Guidelines for Good Practice', ACHCEW, 1986.

'The Consumer's View – A Review of CHC Surveys of Outpatients', ACHCEW, 1986.

Departments

'Subjective Measures of Quality of Life in Britain 1971–1975', Polytechnic of North London, 1976.

'Your Community Health Council in Action', 1984.

'Within Reach of Health Care?', Welsh Consumer Council, 1986.

Health Education Council

'A Guide to Healthy Eating', 1986.

'Beating Heart Disease', 1986.

'Can You Avoid Cancer?', 1985.

'The Health Divide – Inequalities in Health Care in the 1980s', 1987.

'That's the Limit!', 1986.

'The Big Kill – Death and Death Rates from Smoking', 1986.

'So You Want to Stop Smoking?', 1986.

Department of Health and Social Security

'Inequalities in Health Care', Report of a Working Party chaired by Sir Douglas Black (the 'Black Report'), 1979.

'Neighbourhood Nursing – A Focus for Care', Report of the Review of the Community Nursing Service in England (The 'Cumberlege Report'), 1986.

'National Health Service Management – An Inquiry' (the 'Griffiths Report'), 1983.

'Patients First: Consultative Paper on the Structure and Management of the National Health Service in England and Wales', 1979.

'Prevention and Health: Everybody's Business', 1976.

'Priorities for Health and Personal Social Services in England: A Consultative Document', 1976.

'Sharing Resources for Health in England and Wales' (the 'RAWP Report'), 1976.

Leaflet: 'How to Get NHS Treatment', July 1986.

DHSS and the Welsh Office

'Heartbeat Wales', 1987.

Report of the Royal Commission on the National Health Service 1979 (HMSO, Command Paper 7615).

Articles

'Antiquated Hospitals', *Community Health News*, December 1984.

'Lies, Damned Lies and Suppressed Statistics', *British Medical Journal*, August 9th, 1986.

'No Fault Compensation', *Lancet*, June 15th, 1986.

'The Rising Cost of Prescriptions', *New Society*, March 14th, 1986.

'Waiting List Statistics – A Major Deception?' *British Medical Journal*, October 18th, 1986.

'Waiting Lists – Waiting for the Blitz', *Health Services Journal*, August 7th, 1986.

'UK Health All Spent Out: Report on Statistics of Office of Health Economics', *Health Services Journal*, May 22nd, 1986.

Books
Judith Cook, *The Price of Freedom* (New English Library, 1985).
Wendy Farrant, Jill Russell, *The Politics of Health Education* (Bedford Way Papers, 1987).
Innes Pearse, and Lucy D. Crocker, *The Peckham Experiment* (Allen & Unwin, 1943). (See also the further detailed Peckham bibliography under the notes for Chapter 11.)
Patrick Pietroni, *Holistic Living* (Dent, 1986).

Appendix A: The Patients' Charter
PATIENTS' CHARTER
GUIDELINES FOR GOOD PRACTICE

All persons have a right to:

1. health services, appropriate to their needs, regardless of financial means or where they live and without delay;
2. be treated with reasonable skill, care and consideration;
3. written information about health services, including hospitals, community and General Practitioner services;
4. register with a General Practitioner with ease and to be able to change without adverse consequences;
5. be informed about all aspects of their condition and proposed care (including the alternatives available), unless they express a wish to the contrary;
6. accept or refuse treatment (including diagnostic procedures), without affecting the standard of alternative care given;
7. a second opinion;
8. the support of a relative or friend at any time;
9. advocacy and interpreting services;
10. choose whether to participate or not in research trials and be free to withdraw at any time without affecting the standard of alternative care given;
11. only be discharged from hospital after adequate arrangements have been made for their continuing care;
12. privacy for all consultations;
13. be treated at all times with respect for their dignity, personal needs and religious and philosophic beliefs;
14. confidentiality of all records relating to their care;*
15. have access to their own health care records;
16. make a complaint and have it investigated thoroughly, speedily and impartially and be informed of the result;

* Already an established legal right.

17. an independent investigation into all serious medical or other mishaps whilst in NHS care, whether or not a complaint is made, and, where appropriate, adequate redress.

Published by the Association of Community Health Councils for England and Wales.

Appendix B:

Inequalities in Health

From 'The Health Divide – Evidence of Class Differences in Mortality in the Early 1980s', Health Education Council.

Figure 1 illustrates the typical pattern found in early life. In 1984 rates of stillbirth and death at various stages in the first year of life increased progressively from occupational class I to class V. Thus babies whose fathers had unskilled jobs ran approximately twice the risk of stillbirth and death under one year, than babies whose fathers worked in the professions.

Up-to-date statistics on occupational class differences in the death rates of older children will be published by OPCS in April 1987, and these are awaited with great interest as wide differences in health could be seen from the 1970–72 data on this age range.

Evidence of a distinct gradient in death rates in adults of working age is available from two major national sources: the Decennial Supplement for 1979–80, 82–83 and the OPCS Longitudinal Study covering 1976–81 and 1981–83. Figures 2 and 3 illustrate the data from both sources using standardised mortality ratios (SMR).

Figure 1

Outcome of pregnancy by social class of father.
England and Wales, 1984

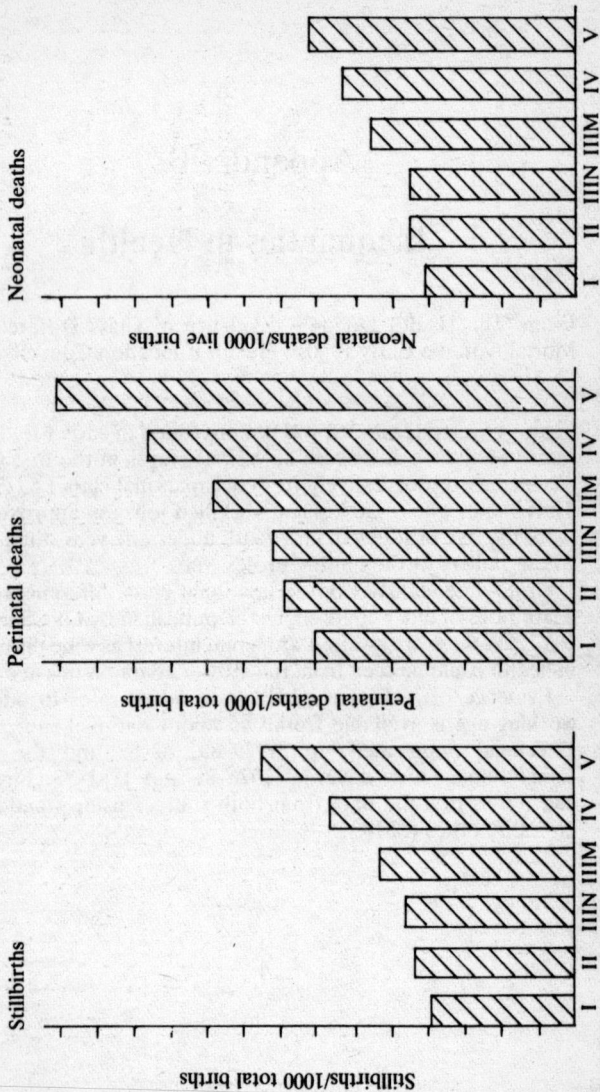

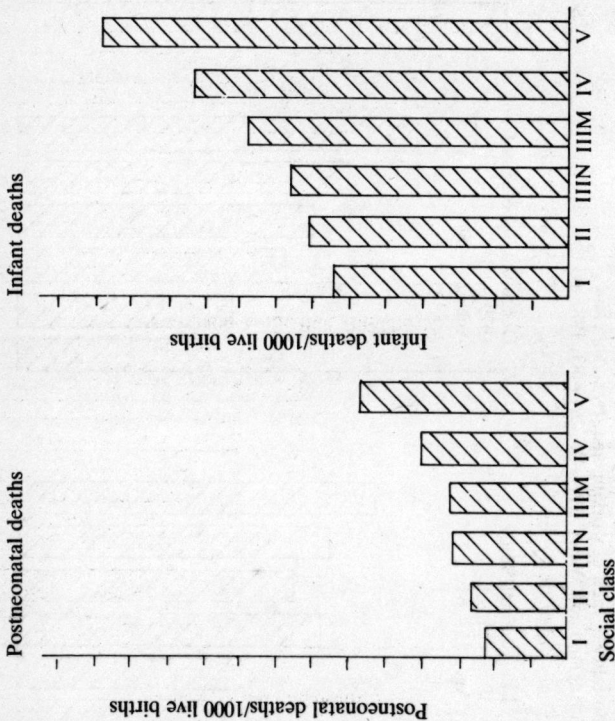

Postneonatal deaths

Postneonatal deaths/1000 live births

Social class

Infant deaths

Infant deaths/1000 live births

Stillbirths:
foetal deaths after
28 completed weeks
of gestation

Perinatal deaths:
stillbirths and
deaths in the first
week of life

Neonatal deaths:
deaths in the first
28 days of life

Postneonatal deaths:
deaths at ages over
28 days and under
one year

Infant deaths:
deaths at all ages
under one year

Source: OPCS (1986).

Figure 2

Mortality of men of working ages by social class
in England and Wales, 1976–81, 1979–83, 1981–83

Source: Drawn from OPCS data.

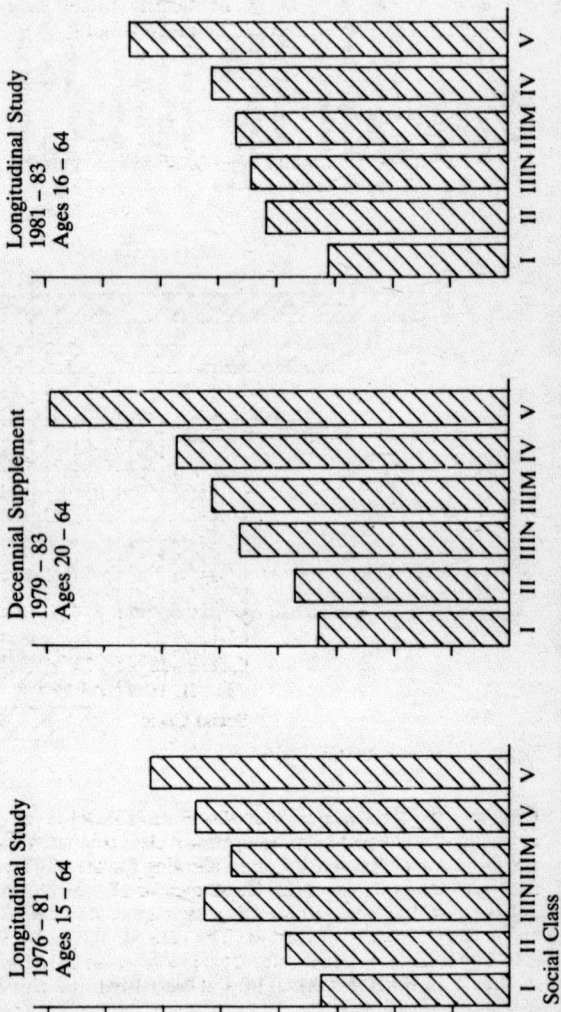

Longitudinal Study
1976 – 81
Ages 15 – 64

Social Class

I   II   IIIN IIIM IV   V

Decennial Supplement
1979 – 83
Ages 20 – 64

I   II   IIIN IIIM IV   V

Longitudinal Study
1981 – 83
Ages 16 – 64

I   II   IIIN IIIM IV   V

Note that the Decennial Supplement records a very high SMR for class V. For technical reasons, related to the re-classification of occupations by the Registrar General in 1980, it is thought that class V figures may be artificially raised and the lower Longitudinal Study figure may be more reliable.

Figure 3

Mortality of women aged 20–59 by social class
in Great Britain, 1979–83

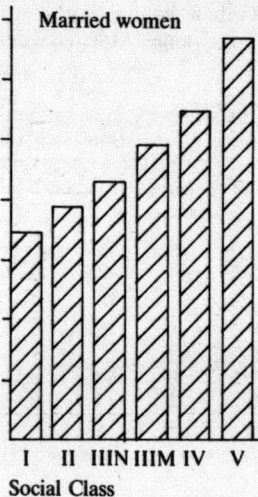

Source: OPCS (1986).

Footnote: the Standardised Mortality Ratio (SMR) is the ratio of the mortality rate in a particular occupational class compared with the average for the whole population, after allowing for the difference between their age structure. The SMR for all men or all women would be 100. SMRs below 100 indicate lower than average death rates. SMRs above 100 indicate higher than average. The data for the early 1980s show a step-wise increase in death rates from low levels in class I to high levels in class V. Unskilled workers run at least twice the risk of death as professionals.

Whose Health is it Anyway?

The mass of information on social class and mortality contained in the Decennial Supplement (DS) can still be used by grouping the manual classes together and comparing them with all the non-manual classes. Alternatively data on class I and II can be combined and then compared with data on combined classes IV and V to compare the health of the richest and poorest classes. Analyses of this nature have already begun. For example, using the Decennial Supplement data, Marmot and McDowall (1986) carried out a careful study of some of the major causes of death for men and married women in the two distinct categories: manual and non-manual. Table 1 shows the results of the study. For all the causes analysed, non-manual workers in 1979–83 had lower SMRs than manual workers. It was particularly striking in the younger age range. Married women showed a similar pattern.

Table 1
– SMRS* FOR SELECT CAUSES OF DEATH IN GREAT BRITAIN 1970–1972 AND 1979–83 AMONG MEN AGED 20–54 AND 55–64

Cause of death	Age	1970–72			1979–83		
		Non-manual	Manual	NM/M	Non-manual	Manual	NM/M
All causes	20–54	98	131	0.75	76	115	0.66
	55–64	99	128	0.77	82	117	0.70
	20–64	99	129	0.77	80	116	0.69
Lung cancer	20–54	93	167	0.56	60	133	0.45
	55–64	85	145	0.59	67	128	0.52
	20–64	87	150	0.58	65	129	0.50
Coronary heart disease	20–54	103	120	0.86	80	113	0.71
	55–64	102	110	0.93	90	117	0.77
	20–64	102	113	0.90	87	114	0.76
Cerebro-vascular disease	20–54	107	141	0.76	73	121	0.60
	55–64	123	151	0.81	77	119	0.65
	20–64	118	148	0.80	76	120	0.63

* For each cause, and within each age group, the SMR for all men in 1979/83 is 100. The 1970–72 figures were standardised with the 1979/83 rates.

Source: Marmot and McDowall (1986).

Townsend et al have also analysed the DS data by cause of death, comparing the two highest and the two lowest occupational classes. The work illustrates that all the major killer diseases now affect the poorest occupational classes more than the rich. In 65 of the 78 disease categories for men SMRs for classes IV and V were higher than for either class I or class II. Only one cause, malignant melanoma, showed the reverse trend. For women 62 of the 82 categories of disease showed higher SMRs for classes IV and V and only 4 showed the reverse. The rest were neutral. Clearly so called 'diseases of affluence' have all but disappeared and what is left is a general health disadvantage of the poor.

Two major longitudinal studies have also shed light on the situation since the publication of the Black Report: the OPCS Longitudinal Study (LS) and the Whitehall Study of civil servants.

The OPCS Longitudinal Study selected a 1% sample of the population of England and Wales from the 1971 census and has continued to follow vital events in the sample and their associated families by use of flagged birth, death and cancer registrations. The study has been able to look at mortality in 1976–81 by occupational class in broad age groups. Figure 4 shows that even in ages after retirement the mortality gradients for men are almost as steep as those found in the later stages of working life. Thus, even in old age, men from occupational class V had more than 50% higher death rates than those in occupational class I. Fox et al (1986) point out that these are 'The first reliable estimates of the social class gradient in England and Wales at these older ages.' The study is currently analysing the data with respect to women.

The Whitehall study of civil servants examined over 17,000 office-based civil servants in London in 1967–69 and has been following the health of the men in different grades of the service for more than a decade now. The study has found that the lower the grade, the higher the mortality for every cause of death except genito-urinary disease. There is a greater than three-fold difference between the highest and the lowest grade: a much steeper gradient than the occupational class gradient found in the national data.

Other studies of single occupations have also found steeper gradients. For example, Lynch and Oelman (1981), looking at mortality from coronary heart disease in different ranks in the

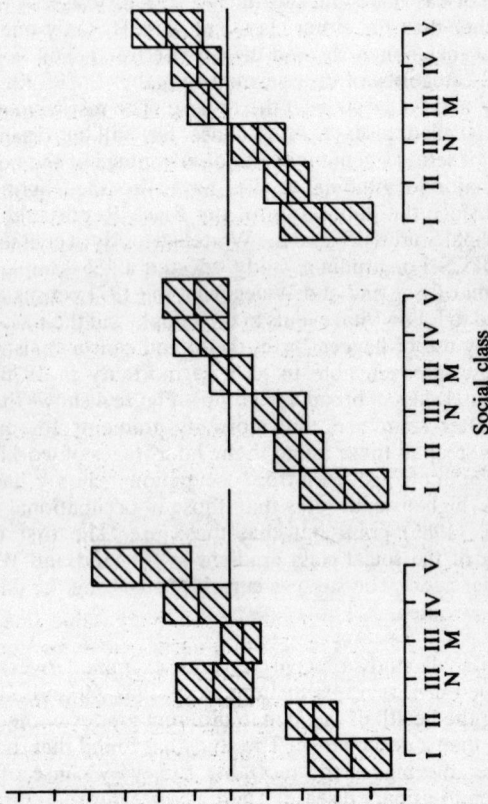

Figure 4

Mortality of men in 1976–81 by social class and broad age groups

Social class

I II IIIN IIIM IV V

hatched areas represent approximate 95% confidence intervals for SMRs; each SMR is indicated by a horizontal line within the hatched area

Source: OPCS Longitudinal Study; Fox et al (1986).

British Army in 1973–77 found increasing mortality with decreasing rank with a five-fold difference between highest and lowest rank.

It seems that the national gradients may underestimate the real differences between social groups, because each class is made up of occupations with a range of mortality rates. Studies on single occupations are more homogeneous and can reveal a sharper distinction between top and bottom groups.

Not only do lower occupational classes have higher death rates, but they also experience more sickness, and ill-health throughout their lives. Indeed Blaxter (1986) argues that morbidity or general health status are more important indicators of inequality now that people are living longer and degenerative disease is becoming more prominent. One of the main sources of information on morbidity is the annual General Household Survey in which people are asked about both chronic and acute sickness. For such purposes six socio-economic groups are used, which correspond fairly well to the Registrar General's scale. The six socio-economic groups are:

1. professional

2. employers and managers

3. intermediate and junior non-manual

4. skilled manual

5. semi-skilled manual and personal service

6. unskilled manual

Rates of sickness have remained relatively stable since 1980. Figure 5 shows the rates for 1984 for each socio-economic group and reveals great inequalities between the groups. The gradient is steepest for limiting long-standing illness, where the rates in the unskilled manual group are more than double those of the professional group for both men and women. The gradient is not quite as steep, but still marked for long-standing illness without limitation; while for acute sickness the differential is only evident at ages over 45.

When age as well as socio-economic group is taken into consideration it is in middle-age that the differences between the socio-economic groups are greatest. See Figure 6.

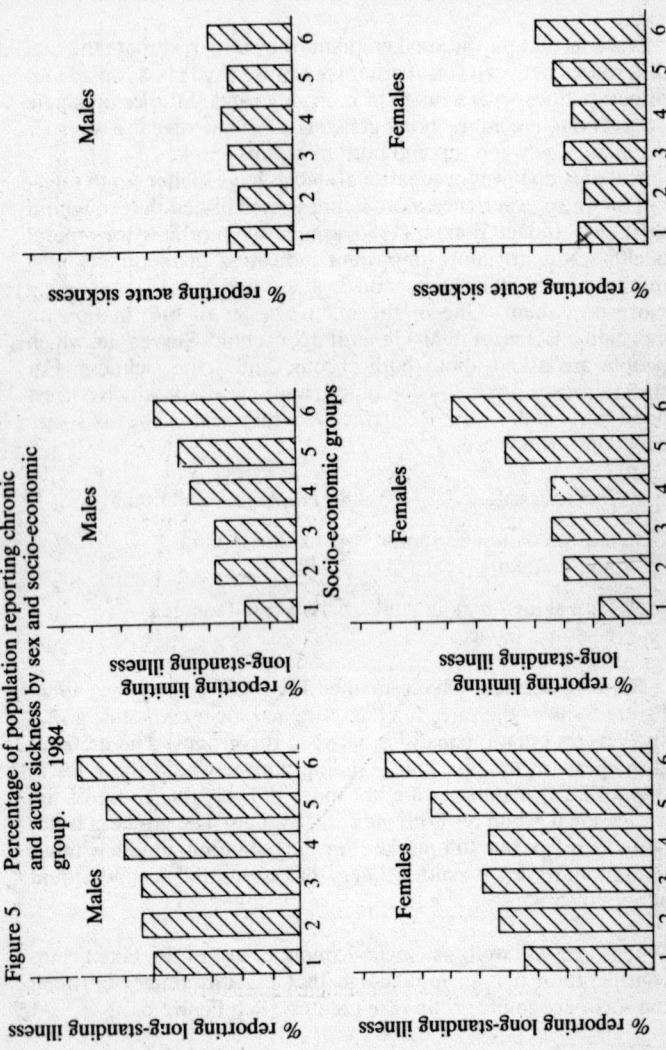

Figure 5 Percentage of population reporting chronic and acute sickness by sex and socio-economic group. 1984

Source: General Household Survey.

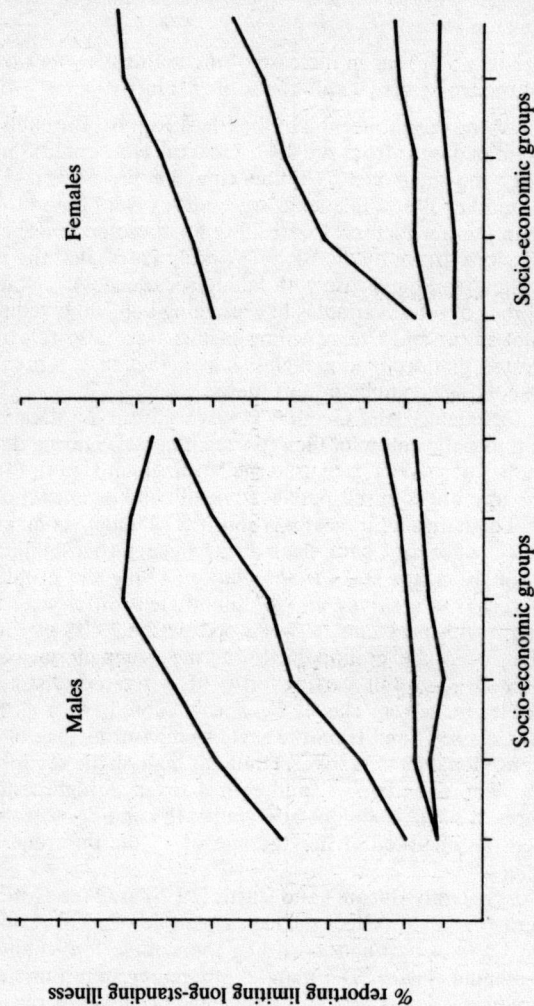

Figure 6   Percentage of population reporting limiting long-standing illness by age and socio-economic group in 1984

% reporting limiting long-standing illness

Males

Females

Socio-economic groups

Socio-economic groups

Source: General Household Survey.

From similar data, Blaxter concludes that health deteriorates more rapidly in those who are socially disadvantaged. She quotes Ferge et al as saying:

'sicknesses appear in the case of the majority sooner or later, but the worse off you are the sooner it hits'.

Following the publication of the Black Report, Burchell (1981) re-examined data from the 1976 General Household Survey to identify the main factors influencing the presence/absence of long-standing illness in adults. Age and sex were found to be the most important factors. Controlling for these he confirmed the conclusions drawn by the Black Working Party: that the rates of long-standing illness rose with falling socio-economic status. As a number of other variables like smoking behaviour, education, marital status and overcrowding factors were also relevant, he suggested that occupational class may be acting as a proxy for a more complex combination of these variables.

Unfortunately, the General Household Survey data give no insight into the nature of the reported illnesses. During the 1980s a number of surveys have used the Nottingham Health Profile to study differences in self-perceived health and its impact on daily lives. The nature of ill-health is considered under six main headings: lack of energy, pain, sleep disturbance, physical immobility, emotional distress and social isolation. Using the profile Hunt et al (1985) in a survey in 1981 found clear differences in perceived health problems between occupational classes, but only in the 20–44 age group: the lower the occupational class the greater the amount and severity of perceived distress. For example, men from classes IV and V scored twice as high as men in classes I and II on the sections measuring lack of energy and emotional distress, over 3 times as high on the section measuring sleep disturbances, and over 4 times as high concerning feelings of social isolation. For women the findings were similar, though less marked. After the age of 45 the differences were smaller.

Most recently Bucquet and Curtis (1986) used the Nottingham Health Profile to collect data on a sample of local residents in three London boroughs, dividing the sample into manual and non-manual classes. The manual groups registered higher rates of tiredness, sleep problems, pain and emotional distress, in line with the findings of Hunt (1985).

In a pilot study of several different measures of health, Blaxter found that manual groups not only had higher rates of chronic disease and disability than non-manual groups, but also were more likely to register a lack of psychological well-being and to have poorer fitness scores.

An important new national study, the Health and Lifestyle Survey, has been investigating other dimensions of health and well-being. The study is unusual in using physiological measurements, like blood pressure and lung function, in addition to in-depth interview techniques and in using income as well as other measures of social disadvantage. Preliminary results have revealed large gradients by social class in a number of conditions like arthritis, haemorrhoids and deafness which can have a considerable effect on the quality of life without threatening life. Differences in self-reported health by income were also found. For instance those with lower incomes tended to report higher rates of illness symptoms and define their own health as poor, compared to their richer counterparts.

Most of the above studies were based on self-reported health but there are numerous studies based on objective medical and dental observations which confirm the overall picture. For example OPCS data for 1984 showed that low birthweight was still much more common in babies born into lower occupational class families. (Low birthweight is considered to be the single most powerful predictor of death in the first month of life.)

In contrast, in the 1980 national survey of heights and weights, childhood and adult obesity was a problem more common in men in class III manual and women in classes IV and V. This was supported by a cohort study of children born in 1946. By 1982 (when the sample was 36 years old) higher rates of obesity were found in lower social groups in both men and women.

Height, which can be taken as an indicator of general health, varies with occupational class. The 1980 survey found that in almost every age group people from manual class households were shorter on average than people from non-manual class households. Marmot, Shipley and Rose (1984) found that height was significantly related to grade in the civil service. For example, the administrators (highest grade) were on average 4.7 cm. taller than men in the 'other' grades (lowest grade).

In the same study, men in lower grades in the civil service had higher blood pressure and lower levels of glucose tolerance than the top grades.

Leon and Wilkinson (1986) examined the survival rates from cancer and coronary heart disease in men and women from the OPCS Longitudinal Study. They concluded that lower class people with cancer had poorer survival prospects.

The adult dental health survey of 1978 found that the proportion of people from class I with no natural teeth was much lower than in the other classes, though class IV and V had shown a marked improvement over the previous decade.

The OPCS survey of children's dental health in 1983 found poorer health in lower occupational classes on several counts. For example, among 5 year olds on average twice as many teeth were actively decayed among those from the manual classes compared to the non-manual classes. They also had more teeth missing due to decay.

References for Appendix B

The Black Report
Blaxter (1986)
Bucquet and Curtis (1986)
Burchell (1981)
The Dicennial Supplement for 1979–80
The Dicennial Supplement for 1982–3
Fox *et al.* (1986)
General Household Survey
Health and Lifestyle Survey
Hunt (1985)
Hunt *et al.* (1985)
Leon and Wilkinson (1986)
Lynch and Oelman (1981)
Marmot and McDowall (1986)
Marmot, Shipley and Rose (1984)
The Nottingham Health Profile
OPCS Longitudinal Study
Townsend *et al.*
The Whitehall Study of Civil Servants

Index

Access to Personal Files Bill, 1986, 108
acid rain, 78
Acre Mill, 76
Advisory Committee on Pesticides, 78
Agent Orange, 79
AIDS (Acquired Immune Deficiency Syndrome), 13, 24, 40, 71, 117
  definition of, 151
  and NHS, 151–2
  progress of disease, 152–3, 166
  transmission of, 153–4
  funding for public information of, 154–5
    for treatment, 155–6
    for research, 156
  drug users and, 157
  media response to, 157
  resources for treating, 158–9
  public reaction to, 160–1, 162–3, 166–7
  doctors' confidentiality and, 162–3
  need for counselling on, 162, 163–4
  precautions against, 164–5
  research into, 165–6
AIDS Action Group, 158
alcohol
  and road accidents, 59, 72–3

related diseases, 72
Allison, Denis (general manager, NW Region), 169
alternative medicine, 141
ambulance service
  for outpatient treatment, 4
  deterioration of, 25, 34–5
  drivers, 25
ancillary services privatisation of, 25, 35–6
ancillary workers, 25
  industrial action (1972–3) by, 17
antibiotics, 13
anti-depressants, 74
Area Health Authorities (AHAs)
  and NHS planning, 14, 15
  abolishing of, 18, 19
Asbestos Information Committee, 76–7
asbestos pollution, 76–7, 80
  related diseases, 113
Association for Improvements in Maternity Services (AIMS), 88, 129, 131
Association of Community Health Councils for England and Wales (ACHCEW), 33, 91, 92
  Services, 101
Association of Radical Midwives (ARM), 105–6

babies
  and AIDS, 154
  stillbirths, neonatal and
    perinatal deaths, 184
  see also childbirth
Bailey, Sir Brian, 67
'Beating Heart Disease'
  (Health Education
  Council), 117
Belgium, 61
Bevan, Aneurin (Minister of
  Health, 1945), 12
Birmingham, 33
  nurse practitioners in, 104
Birmingham Children's
  Hospital, 174
birth defects
  and pesticides, 78
Black, Sir Douglas
  1980 report on inequalities
    in health care, 59, 60, 61,
    65, 67, 68, 69, 70, 75, 99,
    105, 170, 172
Blood Transfusion Service,
  153
  and AIDS, 165
Bloomsbury, 29
Bloomsbury Community
  Health Council, 48
Bloomsbury District Health
  Authority
  AIDS funding of, 155
Body Positive, 161
Bosanquet, Nicholas (Centre
  for Health Economics),
  41
breast cancer deaths, 114
Bristol, 30, 33
Bristol University, 61
British Broadcasting
  Corporation (BBC), 23
British Cardiac Society, 73

British Holistic Medical
  Association, 140
Briti h Medical Association,
  19, 40, 176
British Medical Journal, 51
  on waiting list statistics, 45
  on UK mortality rate, 61
  and Registrar General's
    1986 report, 65
British Nuclear Fuels, 78
Bromley, Kent, 124

Caesarean section, 30
Camberwell, 26
Cambridge, 34
cancer, 25
  and ward closures, 28
  deaths from, 59
  and pesticides, 77
  and nuclear pollution, 78, 79
  see also cervical cancer; lung
    cancer
Cape Industries, 76
Cardiff, 29, 33
'care in the community', 6, 24,
  37–8, 138
  and mentally ill patients, 25,
    37, 55–6
  and old people, 53, 139
  and childbirth, 105
  problems and cost of,
    97–103
Casualty (TV series), 23
Catford, John, 61
Centre for Health Economics,
  41
cervical cancer
  screening, 117, 146
    recall system for, 31–2
  UK deaths from, 59, 114
Chantler, Professor Cyril
  (Acute Unit, Guy's

Hospital), 27
Charing Cross Hospital,
    London, 43
'Charter for Action – Health
    for All . . .' (Faculty of
    Community Medicine),
    58
  on primary health care,
    119–20
chemists *see* pharmacists
childbirth, 124
Child Poverty Action Group,
    69
chiropodists, 96
Chiswick, Dr Malcolm (St
    Mary's Hospital,
    Manchester), 33
Chorley, Lancashire, 34
City Hospital, Truro, 30
  facilities at, 34
City University, London, 62
Clarke, Kenneth (Minister of
    Health, 1984/5), 31
  and drugs prescription, 49,
    50
College of Health, 25
Committee on the Medical
    Aspects of Food Policy
    (COMA), 78
Committee on the Safety of
    Medicines, 78
Community Dental Service,
    110
Community Health Councils
    (CHC), 19, 25, 90–1, 136,
    175
  lay membership of, 12
  and old people's homes, 98
  and primary care Green
    Paper, 106
  and complaints procedure,
    109

community nurses, 111
  and NHS, 47
  and primary care, 96
  and Cumberlege Report,
    103–4
complementary medicine, 145
Confederation of Health
    Service Employees
    (COHSE), 35, 44, 55
Conservative Government
  1951: reviews NHS, 13
  1972: reorganises NHS,
    13–14
  1979: and Royal
    Commission report on
    NHS, 17–18
  and Black Report, 59–60
  1983–7: NHS, 31, 39
    spending on NHS, 24, 40
    and nurses, 42
    and waiting lists, 45, 46
    and drugs expenditure, 49
    rise in prescription and
      dental charges under,
      50–2
    and privatisation of
      optical service, 52
    'care in the community'
      and, 53–5
    unemployment and poor
      health under, 61–2
    economic policy and
      health, 63–4, 69–70
    drug problem and, 74–5,
      157
    health service policies of,
      82, 86–7
    Green Paper (1986) on
      primary health care,
      100–2
    AIDS funding by, 154–7
    and civil service, 169

Conservative Party and NHS, 23, 173
consultants
  and NHS, 12
  and private practice, 17
contraceptive drugs, 52
Cornwall, 34
  health care in, 30, 31, 163, 174
  treatment of cervical cancer in, 32
  waiting lists in, 46
Coventry, 29
  waiting lists in, 46
Crown, Dr June (district medical officer Bloomsbury), 156
Cumberlege, Mrs Julia
  1986 report on nursing service, 100, 103–4, 111
Currie, Edwina (Junior Health Minister), 174, 177

Davidson, Dr Christopher, 169
Denmark, 61
dentists
  and NHS, 12, 47, 96
  and Family Practitioner Committees, 12
  increased charges of, 25, 52
  problems concerning, 110–11
Department of Health and Social Security (DHSS), 18
  and NHS planning, 15
  survey of cervical cancer of, 32
  and waiting list statistics, 45, 47
  cost-cutting circular of, 56–7

suppression of nuclear pollution report, 79
  of 'Health Divide' report, 169
  and centralisation, 139
  figures for AIDS treatment, 155
  and AIDS funding, 159
  response to drug abuse, 157
Department of the Environment, 78
depression, 62
deputising services, 4, 48
Devon, 30
dialysis, 85
diet, 118
  and health problems, 73–4
  and Pioneer Health Centre, 123
  women and, 135–6
dioxins, 78
District Health Authorities, 82
  lay membership of, 12
  reforming of, 19
  London, 25
  and AIDS, 155, 156
  democratisation of, 172
District Management Teams, 18, 21
  and NHS planning, 14
district nurses, 138
doctors *see* consultants; general practitioners; hospital doctors
'domino' system of maternity care, 125
Donaghy, Dr P. (Community Medicine Specialist, Cornwall)
  and AIDS epidemic, 163, 166

Dreadnought Seamen's
Hospital, Greenwich
closure of, 26
Dr Finlay's Casebook (TV
serial), 95
drug abuse, 74–5
government response to,
157–8
AIDS danger to addicts, 157
addiction clinics, 29
drugs
generic prescribing and
'restricted lists', 49–50
see also prescription charges
Duggan, Dr Peter
(psychiatrist, King's
College Hospital), 38
Dulwich Hospital, London, 26
Dunn, Dr Peter (Southmead
Hospital), 33
Dunwoody, Dr John, 26

East Cumbria, 32
Edinburgh, 62
directory of GP's in, 107
AIDS virus in drug users in,
157
Education Act, 1980
and school meals, 65
Environmental Health
Advisory Team, 154
environmental pollution,
76–80
Epsom Hospital, 32
European Economic
Community (EEC), 77,
79
Evening Standard, 27

Faculty of Community
Medicine, 58

Falmouth, 174
family doctors *see* general
practitioners
Family Practitioner
Committees, 12, 48, 102,
130
and NHS reorganisation,
18, 19
and patients' complaints,
95, 108, 109
Farrant, Wendy, 117, 118
Ford, Sherry, 61
Fowler, Norman (Secretary of
State for Social Services),
32
1985/6 NHS report, 24
and Cumberlege Report, 100
response to AIDS, 159
fact-finding tour of Holland,
157
Fox, Professor John, 62
France, 61
AIDS research in, 160, 165

General Medical Council, 129
and complaints procedure,
110
General Medical Services
Committee, 48
1986 report on prescription
charges, 51
general practitioners (GPs)
in inner cities, 4–5
and NHS, 12, 48
problems concerning, 94–5,
107
and old people, 98–9, 139
and 1986 Green Paper on
primary health care,
101–2
and obstetrics, 106
receptionists and, 95, 108

general practitioners *cont.*
  complaints about, 95,
    109–10
  and service to patients,
    129–30, 137
  choice and quality of, 133
  restrictive practices of, 138
  doctor-patient partnership,
    142–4, 145
genetic mutation, 77
Gloucestershire, 30
Great Ormond Street Hospital
    for Sick Children,
    London, 33
Greece, 61
Greenfield Report, 1982, 49,
    50
Gregoire, Dr Alain
    (psychiatrist, King's
    College Hospital), 38
Griffiths, Sir Roy (Chairman
    of Sainsbury's), 20, 21
  1983 report of, 24, 47
Guardian, 23, 37, 43, 152
  survey on premature baby
    care, 32–3
  and health cuts, 168–9
  and 'Register of
    Environmental
    Achievements', 78
Guy, Dr Roland (Charing
    Cross Hospital), 43
Guy's Hospital, London
    closures in, 27, 29
Gwent, 107, 175

Hackney Community Health
    Council, 34
Hammersmith Hospital,
    London, 33
'Health Divide – Inequalities
    in Health . . .' (Health

Education Council,
    1987), 67, 169, 170,
    183–96
health education, 132, 133–5
  new approach to, 143–5
  and AIDS, 152, 154
Health Education Council, 83,
    117, 170
  and Whitehead Report, 67,
    169, 170, 183–96
  disbanding of, 67, 74, 120,
    155
  role of, 120
health visitors
  and NHS, 47
  and primary health care, 96
  and GPs, 138
'Heartbeat Wales' (Welsh
    Office, 1987), 99, 170
heart disease
  deaths in UK, 59
  and diet, 73
heart transplants, 13, 40
  cost of, 85
Hebden Bridge, Yorkshire, 76
heroin, 74
hip replacement operations,
    85
Holistic Living (Dr Patrick
    Pietroni), 140–1, 143
Holland, 61
  childbirth in, 106
  drug users and AIDS in, 157
homosexuals
  and AIDS, 153, 161, 162,
    163, 167
hospital buildings, 11
  updating of, 25, 148
  closure of, 26
  health care in, 40, 41
  location of, 86–7
hospital doctors, 21, 25

and 'pay-bed' dispute (1975), 17
see also consultants
House of Commons Social Services Committee, 102, 176
Hudson, Rock, 165
Human T-lymphotropic virus (type 3) (HTLV3) AIDS virus), 153
Huntingford, Peter, 131
hypothermia, 63
hysterectomies, 30
waiting lists for, 2–3

Income Support Scheme, 64
'Inequalities in Health Care' (Black Report), 60–1, 62, 66, 67, 68, 69, 70, 75, 99, 105, 169, 172
infant mortality in UK, 59
Institute of Health Service Management, 40, 81
report on waiting lists, 46–7
Institut Pasteur, France
AIDS research at, 165
intensive care units, 5
baby care units, 25, 33, 174–5
and nursing shortage, 43
Islington, 115

Jenkin, Patrick (Minister of Health, 1979)
and Black Report, 67–8
Joint Advisory Committee on Nutrition Education (JACNE), 154
Joint Consultative Committees (JCCs), 16
Joseph, Sir Keith (Minister of Health, 1970)

and reorganisation of NHS, 13

Kearns, Dr William (medical officer, NE Thames), 156
kidney patients, 25
kidney transplants, 13
cost of, 85
King's College Hospital, Camberwell
budget cuts at, 26, 29
Department of Psychiatry, 38
King's Fund Centre, 26, 41, 177
Kirkwood, Archie, MP, 108

Labour Government and NHS, 140
1945: sets up NHS, 12–13
1974–9: study in deprivation and health, 59
Labour Party
and NHS, 23, 137, 177
health service policies of, 82–3, 101
and 'winter of discontent' (1978/9), 28
Lang, Dr Tim (Director, London Food Commission), 78
leukaemias, 79
Lewisham Hospital, London, 27
Lewisham/Southwark Health District
1986 crisis in, 27–8
District Health Authority of, 26
'Life before the National Health Service' (TUC East Midlands), 9

Lighthouse, 161
Lisson Grove Health Centre, 140
Liverpool, 33
Liverpool District Health Authority, 29
Liverpool Women's Hospital, 29
liver transplants, 13
Lloyd George, David
and medical insurance for workers (1911), 8
local authorities
and health care, 138–9
London Food Commission, 77
London health care, 46
medical facilities in, 15–16
cuts in, 25–9
premature baby care in, 32–3
Ambulance Service in, 34
psychiatric care in, 37–8
hospital funding in, 41
nurses' housing in, 42–3
waiting lists in, 48
GPs in, 49
AIDS funding in, 159
London Health Emergency, 27
Low Pay Unit, 63
lung cancer
among unemployed, 63
and smoking, 71
and asbestos, 76

Manchester, 115
waiting-list survey in, 45
funding of AIDS treatment in, 156
Maternity Alliance, 43
Maynard, Professor Alan and QUALY units, 85–6

McNair-Wilson, Michael, MP
and 'Patients' Charter', 91
McQueen, Steve, 76
medical accidents, 92–3
Medical Defence Union, 92
medical records, access to, 108
medical training, 132
Medicare (USA), 84
MENCAP, 88
mental illness
closure of hospitals treating, 5–6
care of, 97
and 'care in the community', 25, 37, 56
see also psychiatric hospitals
Menzies, Dr Donald
(consultant, Liverpool Women's Hospital), 29
Merrison, Sir Alec
and Royal Commission on NHS, 17
Merseyside
waiting lists in, 46
midwives
and NHS, 47, 131
and primary health care, 96, 105–6
MIND, 75, 88
Ministry of Agriculture, 78
Moore, John (Health and Social Services Secretary), 177

National Association for the Welfare of Children in Hospital (NAWCH), 88
National Audit Commission 1986 report: 'Care in the community', 55
National Consumer Council (NCC), 33, 34, 109

and primary care Green
Paper, 106, 111
National Health Service
(NHS), 6, 7, 58, 114
setting up of, 8, 9, 11
cost of, 12, 13
and Pioneer Health
Centre, 127–8
reorganisation of, 13–14, 19
reallocation of resources
within, 16, 17
industrial unrest within,
17–18, 173
management within, 20–1,
24, 46–7
cuts in London,
in South West, 25–31
and cervical cancer, 31–2
and premature babies, 32–3
and ambulance and
ancillary services, 34–7
and mentally ill, 37–8
present funding ratios, 39,
40
underfunding, 143,
145–50, 173
future funding, 41, 176
cost-cutting policy in, 36,
168–9
rising costs of, 81
and QUALY, 85–6
nursing shortage in, 42–4
waiting lists in, 45–7
GPs and, 48–9
cost of drugs and
prescription charges,
49–52
dental charges, 52
privatisation of optical
service, 52
'care in the community' and,
53–4

smoking and alcohol abuse
and, 72
bureaucracy in, 82–3
political attitudes to, 81,
82–3
'overview' in, 87
pressure groups and, 87–8
users' rights within, 90–1
medical accidents and, 91–3
primary care and, 94–112
and positive health, 115,
171–2
doctor domination of,
129–31, 142
lack of democracy in,
136–7, 171
GPs' restrictive practices in,
138
and treatment of old people,
139
and AIDS, 151–2, 155–7,
158–9, 163–4
future of, 168, 170–2
National Health Service Act,
1946, 50
Amending Act, 1959, 51
'National Health Service
Management . . .'
(Griffiths Report, DHSS,
1983), 20, 21, 24, 47
National Tranquilliser
Advisory Council
(TRANX), 74, 75
National Union of Public
Employees (NUPE),
44–5
neighbourhood nursing service
(NNS), 103
New Cross Hospital, London,
26
New End Hospital, London
closure of, 26

New Society, 53, 127
Newton, Tony (Health
    Minister), 176, 177
New Zealand, 92
Northallerton, 89
North East Thames Region,
    156
North Tyneside, 32
Northumberland, 32
North West Durham, 32
North West Region Health
    Authority, 169
Norton-Taylor, Richard, 79,
    169
Norway, 61
Norwich, 29
Nuffield Centre for Health
    Service Studies, Leeds, 62
nurse practitioners, 104, 111
    success of, 104–5
nurses, 24
    pay dispute of (1974), 17
    and general management,
        21
    statistics of numbers
        employed, 42
    payment of, 42
    shortage of, 29, 43, 147, 174
    psychiatric, 44
    and Cumberlege Report,
        103
    homes, selling of, 42
Nursing Council and
    complaints procedure,
    109–10

Observer, 29, 70, 155
occupational therapists, 47
Office of Health Economics
    1986 report: waiting-list
        figures, 44

Office of Population Censuses
    and Surveys, 62
old people
    care of, 39
    rise in numbers of, 40, 41
    and 'care in the community',
        53, 97, 98–9, 139
    residential homes, 98–9
opticians, 96
    and Family Practitioner
        Committees, 12
    and NHS, 11, 48
    privatisation of, 52, 111
Organisation of Economic
    Co-operation and
    Development, 39
outpatient facilities, 1–2
Oxford, 137, 138
Oxfordshire Community
    Health Council, 31, 129,
    136

Padstow, Cornwall, 28
panel patients, 8, 9
Parker, Richard (St John's
    Hospital, Lincoln), 55
patient-doctor partnership,
    141–3, 144
'Patient Insurance Scheme', 92
Patients' Association, 25, 88,
    89, 129, 132
Patients' Charter, 91
'Patients First . . .' (DHSS,
    1979), 19
Patients' Liaison Committee,
    108
Pearse, Dr Innes
    and Peckham Experiment,
        122, 125, 126, 171–2
Peckham Experiment, The
    (Innes Pearse), 125, 126

see also Pioneer Health
    Centre, Peckham
Penzance, 174
Perrett, Dr Tony (Royal
    Cornwall Hospital)
  and NHS funding, 145–7
  and poverty and ill health,
    149
  and lack of patients
    complaining, 149–50
pesticides, 77–8, 79
pharmacists, 96
  and Family Practitioner
    Committees, 12
  and NHS, 48
  and primary care Green
    Paper, 111
  see also prescription charges
Phillips, Aileen (Director of
    Nursing, Charing Cross
    Hospital), 43
physiotherapists, 96
Pietroni, Dr Patrick, 171
  and holistic approach to
    health care, 140–1
  and doctor-patient
    relationship, 141–4
Pioneer Health Centre,
    Peckham
  work of, 123–6, 138, 143,
    171, 172
  and NHS, 127–8
Plymouth, 30, 31, 163
*Politics of Health Education,
    The* (Wendy Farrant, Jill
    Russell), 117, 120
premature baby units
  and QUALY, 86
prescription charges, 11
  increases in, 25, 50–2, 53
'Prevention and Health . . .'
    (DHSS, 1976), 15

preventive medicine, 15
primary health care, 170
  GPs and, 94–5, 107
  dentists and, 96, 110
  and 'care in the community',
    97–9
    old people and, 97–8
  poverty and, 99
  1986 Green Paper on,
    100–3, 106, 111
  nurse practitioners and,
    104–5
  midwives and, 105–6
  complaints procedure in,
    109–10
  opticians and pharmacists
    in, 110–11
  WHO and, 119–20
  new approach to, 140–2
'Priorities for Health and
    Personal Social Services
    . . .' (DHSS, 1976), 15
private consultations, 3
private health insurance, 84
private practice and NHS,
    11–12, 17
psychiatric hospitals
  and nursing shortage, 43–7
  management of, 47
  see also mental illness
Pulse magazine, 32

QUALY ('Quality Adjusted
    Life Years') units, 85–6
Queen Mary's Hospital for
    Children, Carshalton
  proposed closure of, 27, 33

radioactive waste, 78
'Red Alert' emergency
    admissions, 29
  at Treliske Hospital, 30

Regional Health Authorities,
    18, 19, 82
  lay membership of, 12
  budgets of, 16
  and responsibility for NHS,
    18
  London, 25
  and smear-test recalls, 31–2
  and GPs, 48–9
  and CHC, 90
  and centralisation, 139
  democratisation of, 171
Registrar General
  1986 analysis of UK deaths,
    65, 134–5
Resource Allocation Working
    Party (RAWP), 30
  1976 report of, 16, 17
Richardson, Tom
    (Oxfordshire Community
    Health Council)
  on NHS problems, 136–40
Riverside Community Health
    Council
  and privatised ancillary
    services, 35–6
Robinson, Jean (General
    Medical Council, 171
  on NHS problems, 129–36
Rochdale health cuts, 169
Roof, SHELTER magazine,
    55
Royal College of General
    Practitioners, 132
Royal College of Nursing, 40
Royal College of Physicians,
    71
Royal College of Surgeons,
    176
Royal Commission on the
    NHS, 1979 report, 17–18,
    19

Royal Berkshire Hospital,
    Reading, 174
Royal Cornwall Hospital,
    Truro, 145
Royal Free Hospital, London,
    41, 122
Royal Sussex and County
    Hospital, Brighton, 175
Russell, Jill, 117

St Bartholomew's Hospital,
    London, 41
St James Hospital, Balham, 26
St John's Hospital, Lincoln
  suicides at, 55
St John's Skin Hospital,
    London
  closure of, 26, 29
St Mary's Hospital, London,
    26
St Mary's Hospital,
    Manchester, 33
St Thomas's Hospital,
    Lambeth, 41
  ward closures in, 27–8, 175
Savage, Wendy, 131
Scarborough Hospital, 175
school meals, 65, 70
Scotland
  deaths from heart disease,
    61
Scottish Consumer Council,
    107
Scott Williamson, Dr George
  and Peckham experiment,
    122
SDP-Liberal Alliance, 66, 177
  health service policies of, 83
Sellafield, 78
Sheffield, 99, 117
  Survey, 170

Silverstone, Dr Peter (psychiatrist, King's College Hospital), 38
smear tests
recall system for, 31–2
see also cervical cancer
smoking, 99
and health problems, 71–2, 134–5
and women, 117, 134–5
Snow, Dr John, 113
Social Science Research Council, 118
Social Security Act, 1986, 64
South East Thames Health Region, 156
South East Thames Regional Health Authority, 42
Southmead Hospital, Bristol, 33
South Tees, 32
South West Regional Health Authority, 30
Spastics Society, 88
Spearing, Nigel, MP, 110
speech therapists
and NHS, 47
and primary health care, 96
State Earnings Related Pension, 64
Stockton, Lord, 66
Sunderland, 32
Sweden, 61
patients' compensation scheme in, 92
spending on health, 39

Tax cuts, 177
Taylor, Matthew, MP, 174
Tebbit, Norman (Chairman of Conservative Party)
and NHS cuts, 23

Terrence Higgins Trust, 161, 163
Thatcher, Denis, 35
Thatcher, Margaret, 18, 28, 173, 176, 177
and NHS, 23, 31
government of, 46, 50, 60, 62, 67, 74, 82
and relation between unemployment and poor health, 62
response to drug trafficking, 157
This Week (TV programme), 86
report on health in Sheffield, 99, 117
Tooting Bec Hospital, 29
Townsend, Professor Peter, 60, 61, 66
tranquillisers, 74–6
transplant operations, 13
cost of, 85
Treliske District General Hospital, Truro, 30
Trent, 55
Trent Regional Health Authority, 44, 47
Truro, 30
TUC, East Midland Region
'Life before the National Health Service', 9–11

Uganda
AIDS in, 152
UKCCN Act, 110
unemployment
and poor health, 62–3, 99
United States of America, 165
spending on health, 39

United States of America –
    cont.
  private health insurance in,
    84
  AIDS in, 152, 153, 154
    research into, 156, 160
University College Hospital,
    London, 33
US National Institute of
    Health, 165

waiting lists, 25, 27
  in Lewisham Hospital, 26–7
  in London, 29
  rise of, 44–5
  and readmissions, 47
Wales
  NHS in, 170
Wandle Valley Hospital,
    Merton, 26
Ward, Colin, 127, 128
ward rounds, 89
water pollution, 77
Wates St Mary's Health Care
    Research Unit, 140
Wellcome Foundation, 165
Well Woman Centres, 115–16
West Germany, 61
  spending on health, 39
West Lambeth District Health
    Authority
  1986 cuts by, 27, 28
West Middlesex Hospital,
    London, 33
West Midlands Health
    Management Centre

waiting-list survey of, 46
Westminster Children's
    Hospital, London, 26
Westminster Hospital,
    London
  privatised ancillary services
    at, 35–6
Whitehead, Margaret
  1987 report on health care
    inequalities, 67, 68, 69,
    70, 183–96
Whitelaw, Dr Andrew
    (Hammersmith Hospital),
    33
Williams, Dr Roger
    (consultant, King's
    College Hospital), 26
Williams, Sir Owen, 123
Winterton, Dr Michael
    (consultant, Treliske
    Hospital), 30
women
  increase of smokers among,
    71–2, 134–5
  alcohol and, 72
  health screening of,
    114–15
  diet and, 135–6
  and AIDS, 153
Woodhouse, David, 92
World Health Organisation,
    58, 85, 119

York University, Centre for
    Health Economics, 85